Advances in Anatomy
Embryology and Cell Biology

Vol. 68

Editors
A. Brodal, Oslo   W. Hild, Galveston
J. van Limborgh, Amsterdam
R. Ortmann, Köln   J.E. Pauly, Little Rock
T.H. Schiebler, Würzburg   E. Wolff, Paris

Agnes A.M. Gribnau
Leonardus G.M. Geijsberts

Developmental Stages in the Rhesus Monkey (Macaca mulatta)

With 27 Figures

Springer-Verlag
Berlin Heidelberg New York 1981

Dr. Agnes A.M. Gribnau
Leonardus G.M. Geijsberts
Department of Anatomy and Embryology
University of Nijmegen
P.O. Box 9101
6500 HB Nijmegen
The Netherlands

ISBN-13:978-3-540-10469-8   e-ISBN-13:978-3-642-67908-7
DOI: 10.1007/978-3-642-67908-7

Library of Congress Cataloging in Publication Data.
Gribnau, Agnes Antoinette Marie. Developmental stages in the rhesus
monkey (Macaca mulatta). (Advances in anatomy, embryology, and
cell biology; v. 68) Bibliography: p. Includes index. 1. Rhesus monkey-
-Development. 2. Embryology--Mammals. 3. Mammals--Development.
I. Geijsberts, L. G. M., 1927- joint author. II. Title. III. Series. [DNLM:
1. MACACA--Embryology. 2. MACACA--Growth and development.
Wl AD433K v. 68/QL 737.P93 G846d]
QL801.E67 vol. 68 [QL959] 574.4s [599.8'2] ISBN-13:978-3-540-10
469-8

Composition: Schreibsatz-Service Weihrauch, Würzburg

2121/3321-543210

Contents

Acknowledgments

The authors wish to thank Dr. H.J. ten Donkelaar and Prof. Dr. R. Nieuwenhuys for critical reading of the manuscript, Mrs. Y. van den Bosch-Willemse and Mr. H. van Aanholt for their histo-technical assistance, and Miss M. Sjak Shie for typing the manuscript.

The maintenance, breeding and surgery of the animals was skilfully conducted by Mr. A.M. Peters, Dr. J.P. Koopman, and Mr. G.J.F. Grutters at the Central Animal Laboratory of the Medical School of the University of Nijmegen.

1 Introduction

In the past decennia nonhuman primates have been increasingly used for research purposes in various scientific fields. Much interest has been focused on this group of animals in general and on the rhesus monkey in particular because of its close phylogenetic relationship with man. In some fields of research, however, such as embryology and microscopic anatomy, much less attention has been paid to nonhuman primates, probably because of the expense involved in the collection of the extensive material needed. On the other hand, teratological and experimental embryologic studies must be based upon a thorough knowledge of the normal ontogenesis since only in that way can a reliable distinction be made between normal and abnormal or induced development.

Each ontogenetic study essentially consists of a comparison of different developmental stages. In most reports dealing with the development of individual organs or their subunits the material used is classified according to the estimated age or the length of the embryos. These criteria, however, are not valid, since considerable variation in developmental stage occurs between animals of the same age even between littermates and between animals of the same length. Therefore a method is needed for assigning embryos to successive developmental stages that are defined on the basis of external and internal characteristics. This type of classification was elaborated by Streeter (1942, 1945, 1948, 1951), who arranged human embryos into developmental horizons numbered XI through XXIII. The preceding horizons I to X were subsequently defined by other authors (Heuser and Streeter 1941; Hertig et al. 1956; Heuser and Corner 1957; O'Rahilly 1973a).

Apart from the age and size of the embryos Streeter used other criteria to characterize the horizons that were derived from the external form, particularly the number of somites and the developmental state of the limbs on the one hand, and from the development of several easily recognizable inner structures and organ systems, such as the eyes, the ears, the intestinal tract, the urogenital, vascular, and central nervous systems, on the other. For convenience O'Rahilly (1972, 1973a) replaced the term 'horizon' by stage and the Roman numerals by Arabic numbers. Unfortunately, Streeter defined his horizons as age groups ranging from 24 ± 1 days in horizon XI up to 47 ± 1 days in horizon XXIII, in which each horizon was 2 days apart. On the basis of data of other authors (e.g., Olivier and Pineau 1962; Nishimura et al. 1968) O'Rahilly also revised the respective ovulation ages of the various developmental stages. In this way a standard system of staging human embryos based upon Streeter's system was introduced, which is currently used by O'Rahilly (1963–1979) together with his co-workers (1956–1979) in the Carnegie Institution of Washington to describe the development of various organs. Streeter's staging system was also employed as a basis for clinically oriented textbooks on the developing human (Moore 1973) and on the normal and abnormal development of the human nervous system (Lemire et al. 1975).

During prenatal development four consecutive periods can be discerned which on the basis of Streeter's developmental horizons in human embryos (1942, 1945, 1948, 1951), as modified by O'Rahilly (1972, 1973a, 1978, 1979) and Hendrickx (1971),

1

comprise the following stages: (a) the presomite stages 1–8 (b) the somite stages 9–12, (c) the postsomite or organogenetic stages 13–23, and (d) the fetal stages. The end of the embryonic period and thus the beginning of the fetal period was arbitrarily defined by Streeter at the end of horizon XXIII as the moment of bone marrow invasion into the cartilaginous precursor of the humerus. Hendrickx (1971) and his co-workers demonstrated, however, that both the timing and the mechanism of palate closure are very similar in man and nonhuman primates. They introduced, therefore, the closure of the secondary palate as the point of transition from embryo to fetus at the end of stage 23, as was also suggested by Wilson (1973), and recently adopted by O'Rahilly (1978).

In a comprehensive study Hendrickx (1971) demonstrated that embryos of the baboon (*Papio cynocephalus*) could also be arranged into 23 successive developmental stages, which on the basis of similar external and internal characteristics correspond to Streeter's horizons in human embryos. Subsequently Hendrickx (1972a, b) Hendrickx and Houston (1971), Hendrickx and Sawyer (1975) and Hendrickx et al. (1975), in a comparative analysis of the embryogenesis in the baboon, rhesus monkey (*Macaca mulatta*), bonnet monkey (*Macaca arctoides*), cynomolgus monkey (*Macaca fascicularis*), green monkey (*Cercopithecus sp.*), and greater bush baby (*Galago crassicaudatus*), concluded that the embryonic development of the nonhuman primates is essentially similar. The same holds for the lesser galago (*Galago senegalensis*) as demonstrated by Butler (1972) and probably also for a primitive primate (*Tupaia belangeri*) as analyzed by Kuhn and Schwaier (1973). The description of comparable embryonic stages 7–23 in the common marmoset (*Callithrix jacchus*) by Phillips (1976) proves that the 23 developmental stages originally described by Streeter (1942, 1945, 1948, 1951) in human embryos are not only applicable to Old World monkeys and prosimians but also to New World monkeys. In general, however, all the reports mentioned lead to the conclusion that considerable differences occur in the timetable of the different developmental stages. The most conspicuous feature is the clear delay in development in the galago and the common marmoset as can be seen in Table 1. In addition species-

Table 1. Embryonic age of developmental stages in some primate species

Developmental stage	Age in days			
	Man (O'Rahilly, 1979)	Baboon (Hendrickx, 1971)	Galago (Butler, 1972)	Marmoset (Phillips, 1976)
7	±16	16–18	±25	25–28
13	±28	28–30	±38	±65
17	±41	34–36		±75
23	±57	46–48	±60	±83

specific anatomic and temporal differences exist in the development of the individual organs.

Only a few studies on the embryogenesis of the rhesus monkey are available. Heuser and Streeter (1941) correlated the estimated gestational age of a series of embryos with their external form and size, dealing primarily with early embryonic stages, whereas in an earlier paper Wislocki and Streeter (1938) described the events of placentation. Fetal growth and development of the rhesus monkey were analyzed by Schultz (1937), van Wagenen et al. (1965), and Kerr et al. (1972), whereas Tanimura and Tanioka (1975) made a comparison of embryonic and fetal development in man and rhesus monkey. Most of the more recent reports on the development of the rhesus monkey deal with specific organogenetic features such as ossification (van Wagenen and Asling 1964; Ferron et al. 1976), the gonads (van Wagenen and Simpson 1965), and the palate (Asling and van Wagenen 1967; Steffek et al. 1968). Hendrickx and Sawyer (1975) reviewed the available information on the embryology of the rhesus monkey and provided a general account of the prenatal development of this species. These authors also stated, however, that in many instances the results were based upon limited observations and that much more research was necessary to understand how normal development proceeds. This is especially true for the organogenetic period on which only a short checklist of criteria was presented.

Thus the first aim of the present study is to fill this gap by giving a detailed description of the organogenetic developmental stages 13 through 23 in the rhesus monkey conforming with those described by Streeter. The staging of the embryos is executed on the basis of three sets of criteria: (1) data on the postconceptional age and size of the embryos; (2) external characteristics, such as outer form and developmental state of outer structures; and (3) internal characteristics derived from microscopical sections. The most appropriate structures to be used in the staging procedure are the limbs, the eye, and the ear, and other structures will be also included in the description, for instance the central nervous system etc. But it should be noted that the development of the digestive, circulatory, and urogenital systems will not be considered since these systems are too complex to be used in a staging procedure.

The detailed studies of Rugh (1968) and Theiler (1972) on the mouse provide evidence that in rodents certain features of the organogenesis can similarly be used to determine the classification of embryos and fetuses into developmental stages. Theiler roughly related the various stages in the development of the mouse with the developmental horizons of Streeter. Some phases in the development of the mouse appeared to be faster than the corresponding phases in man, whereas others turned out to be slower. Similar results were recently described by ten Donkelaar et al. (1979) for the Chinese hamster (*Cricetulus griseus*). These authors used a system of staging more in conformity with that of Streeter and in this way they were able to demonstrate a good concordance in development between primates including man and Chinese hamster, at least up to stage 17. Apart from the substantial evidence provided by the reports mentioned the rather brief studies of Scott (1973) on the guinea pig, Graves (1945) on the golden hamster, Otis and Brent (1954) on mouse and man, Christie (1964) on the rat, and Edwards (1968) on the rabbit and rat also suggest that not only other primates but also rodents follow a common, basic developmental pattern similar to that of man.

Thus the second objective of the present study is to analyze whether the staging system used in the rhesus monkey and essentially based upon Streeter's horizons is applicable not only to other primates but also to rodents. If so, it would not only put an end to the very confusing numbering of the successive developmental stages as introduced by the respective authors for various species, but also allow a direct comparison of the various mammalian species in different phases of their embryonic development.

2 Materials and Methods

2.1 Animals

A rhesus monkey (*Macaca mulatta*) colony is maintained at the Central Animal Laboratory of the University of Nijmegen. The population used in the present study consisted on average of 20 adult females and 3 adult males, placed in individual cages. During the whole project a total of 64 females and 8 males were used. The animals were fed on a commercially prepared pellet diet (Hope Farms B.V. Woerden) with additional fruit and water ad libitum.

Each female was examined daily for vaginal bleeding to record the menstrual cycle, which in most specimens lasts 28 days, but in many cases varies between 27 and 31 days. In this way the implantation bleedings that occur in some 80% of the pregnant females are also recorded. This type of bleeding starts at about the 20th day after conception and may last 2–10 days. On average the gestation period in our population lasts approximately 168 days.

2.2 Breeding Method and Pregnancy Diagnosis

At the time selected for mating the female specimen was individually exposed to one male. Since in the rhesus monkey the ovulation occurs somewhere between day 11 and day 14 of the menstrual cycle, it was decided to cage the female with the male during a 72-h period starting on the morning of day 10 after the onset of menstrual bleeding. This 3-day period was a compromise between the longest possible period to effect pregnancy in as many cases as possible and the shortest possible period to reduce the inexactitude of the embryonic age (see below). In this way 30% of all matings resulted in pregnancy; most conceptions occurred in the period between October and May each year. Of all pregnancies diagnosed in our population about 19% spontaneously aborted.

Pregnancy was diagnosed on the basis of immunologic detection of Macaca Chorionic Gonadotropin (MCG) in the urine of pregnant animals. Two different pregnancy tests were used: (1) the immunologic rhesus monkey pregnancy test as described previously (Gribnau 1975) and (2) the 'subhuman primate tube test for pregnancy', which was a gift from the Center for Population Research NICHD by Ortho Diagnostics Inc., Raritan, New Jersey 08869. Both tests were executed on postmating days 19, 20, and 21 since the MCG concentration in the urine of pregnant rhesus monkeys peaks between day 19 and day 25, as demonstrated by Tullner (1968). The first test is rather time-consuming, as it starts with 24-h urine samples, the second test is much more convenient since only a few drops of urine are needed. From our experience it can be concluded, however, that the former method of testing is much more reliable: the false-negative rates being about 10% and 45% respectively. Both test methods showed a false-postive rate of about 3%.

2.3 Estimation of Embryonic Age

The embryonic age was calculated by counting the number of days (x) elapsed after the second day of mating. The latter day was considered to be postconceptional day 1 or embryonic day 1 (E_1). The error in the embryonic age resulting from the mating procedure amounts to $+/- 1$ day. Thus the age of the embryos will be indicated as $E_{x\pm1}$. In our analysis of the whole organogenetic period embryos ranging in age from $E_{28\pm1}$ to $E_{50\pm1}$ were used.

2.4 Hysterotomy Technique

Hysterotomy was performed by a standard surgical technique. At the selected duration of pregnancy the pregnant female, which on the average weighs 4–5 kg, received a single injection of 10 mg Ketalar/kg body wt. together with 0.25 mg atropine. The operation was carried out under deep O_2, N_2O, halothane anesthesia.

The abdominal wall and the peritoneum were opened by a midventral incision and the gravid uterus exposed manually. A horizontal incision was then made in the ventral uterine wall between the two main branches of the myometrial vascular tree. The chorionic sac was dissected out intact, great care being taken not to puncture the chorion. The sac was removed by using a spatula to separate the placenta and basal decidual layer from the parietal decidual layer of the endometrium. After removal of the chorionic sac an intramuscular injection of 0.2 ml oxytocin (10 IU/ml) was given into the uterus to accelerate its contraction. The uterus was then closed with sutures of 3/0 chromic catgut. After the administration of about 20 ml Rheomacrodex, Organon, The Netherlands (10% dextran 40, 5% glucose in 0.9% NaCl) into the peritoneal cavity to reduce adhesion between uterus, bladder and/or omentum majus. The peritoneum and the abdominal muscles were sutured separately with 2/0 chromic catgut. The skin was then closed with linen thread.

The total duration of the anesthesia usually amounted to 75–90 min. The skin sutures were invariably removed by the monkey within 2 weeks. After surgery the female was allowed to recover for about 2 months. Thereafter she was mated again and in the case of subsequent pregnancy reoperated upon. This procedure could be repeated in the same female 4–8 times, depending on the amount of connective tissue and/or adhesions formed within the uterus wall and between the uterus and its adnexa. These were both found to reduce the fertility of the animals.

2.5 Collection and Processing of the Embryos

Immediately after removal from the uterus the chorionic sac was placed in a glass dish and immersed in isotonic physiologic saline. The chorionic sac was then halved, exposing the amniotic sac together with the yolk sac. Subsequently, the transparent amniotic sac was opened and the umbilical cord pinched off and cut. The embryo was then immediately transferred to the fixative consisting of Bouin's fluid (Romeis 1968). As was established empirically, this fixative causes minimum shrinkage in embryonic material.

After fixation for at least 48 h the specimens were processed in alcohol (progressively ranging from 50% to 100%), methylbenzoate, amylacetate and paraffin wax (melting point 56 °C) according to a time schedule adapted to the size of the material. All embryos were serially sectioned on a Spencer rotary microtome at 7 μm in one of the three conventional directions. The sectioning occurred in such a way that for each developmental stage at least both a horizontally and a frontally sectioned series were available. For detailed information on the orientation of the material during embedding as well as the making of reconstructions the reader is referred to an earlier publication (Gribnau and Lammers 1976). The sections were stained either with hematoxylin and eosin or according to Mayer's modification of the Nissl technique (Romeis 1968).

2.6 Photography

All embryos were photographed both directly after fixation in the first alcohol phase (50%) and during the methylbenzoate phase during which they become translucent. In the former photographs a millimeter scale was also included to record the size of the embryos. Small embryos were photographed under a Zeiss operation microscope with a Zeiss/Ikon camera unit, whereas gross specimens were photographed with a Thagee EXA 500 camera.

Microscopic sections were photographed with either an automatic Zeiss Photomikroscop II or a Leitz Aristophot.

Ilford Pan F film was used in all cases except for the Aristophot pictures, for which the Agfa Pan 25 Prof film was more suitable.

3 Description of the Stages

The description of each developmental stage will consecutively deal with: (1) the postconceptional age and the length of the rhesus monkey embryos representative of that stage, (2) the external features of those embryos, (3) their internal features, and (4) a comparison with the embryos of the corresponding stage in other primates and rodents.

The embryos were assigned to stages on the basis of their external and internal characteristics essentially according to the method introduced by Streeter (1942, 1945, 1948, 1951) for man. In horizons XIX through XXIII, however, Streeter (1951) used a method of rating the developmental state of a number of key organs on a system of point scores. The organs selected by Streeter were the cornea, the optic nerve, the cochlea, the hypophysis, the vomeronasal organ, the submandibular gland, the kidney, and the skeleton. In his description of the development of the baboon Hendrickx (1971) consequently followed Streeter's original descriptive approach up to stage 23 inclusive. In stages 19 through 23 Hendrickx additionally described the development of the palate and adjacent areas; closure of the secondary palate marks the end of stage 23. This method was followed by Phillips (1976) in the marmoset, by Theiler (1972) in the mouse, and by ten Donkelaar et al. (1979) in the Chinese hamster, although in the rodents the classification of stages 18–23 appeared to be difficult. The method used also in the present investigation is thus essentially based upon Streeter's procedure as modified by Hendrickx (1971), but also takes account of the developmental state of the central nervous system.

The closure of the secondary palate is a rather complex process, but deserves some general comments since it plays an essential role in the subdivision of the stages 20—23. Within the oronasal cavity the two lateral palatine processes, present in stage 19, show a medially directed outgrowth as well as a rotation from lateral to medial. As a result the lateral palatine processes, which originally are oriented bilaterally to the developing tongue, move toward a horizontal position above the tongue. Anteriorly, the formation of the secondary palate is effected by the union of the palatine processes with the primary palate and the downward growing nasal septum. Posteriorly, the closure of the palate is achieved by the fusion in the midline of the palatine processes with each other and with the nasal septum. The entire process as described shows an anteroposterior developmental gradient. For more detailed information the reader is referred to the studies of Bollert and Hendrickx (1971), Burdi and Faist (1967), Coleman (1965), Fulton (1957), Greene and Pratt (1976), Hughes et al. (1967), Koch and Smiley (1973), Smiley and Koch (1971), Walker and Fraser (1956), and Zeiler et al. (1964) among others.

As we have mentioned, in the description of each developmental stage the fourth section will deal with a comparison of the results found in the rhesus monkey with data on other primates, including man, and rodents. The data on the species involved (man, baboon, marmoset, Chinese hamster, and mouse) are derived from the investigations on the developmental stages by the authors named above. The extensive studies of O'Rahilly and his co-workers (1966—1979) on the staged development of man will also be taken into account. Additional information is derived from the studies of Butler (1972) on the lesser galago, Scott (1937) on the guinea pig, Edwards (1968) on the rabbit and rat, and Christie (1964) on the rat.

3.1 Stage 13

3.1.1 Postconceptional Age and Length

The postconceptional age of the five rhesus monkey embryos examined that were representative of stage 13 varied from 28 ± 1 to 30 ± 1 days. One specimen aged 32 ± 1 days postconception (p.c.) was found that also belonged to stage 13. The greatest length of the stage 13 embryos ranged from 4.5 to 6.0 mm.

3.1.2 External Features

In stage 13 embryos the posterior neuropore is closed since the process of closure of the neural tube is completed during stage 12. The outer form of the embryos is dominated by their C-shaped body axis with apparent thoracic and sacral curvatures (Fig. 1A and B). As a consequence, the heart bulge is almost in contact with the head. The form of the head is mainly determined by the central nervous system (CNS) as can be seen in embryos cleared in methylbenzoate (Fig. 1B). Three branchial bars are present, the first and second of which are very prominent. The first branchial bar has started to subdivide; cranial to the mandibular process a small maxillary process can

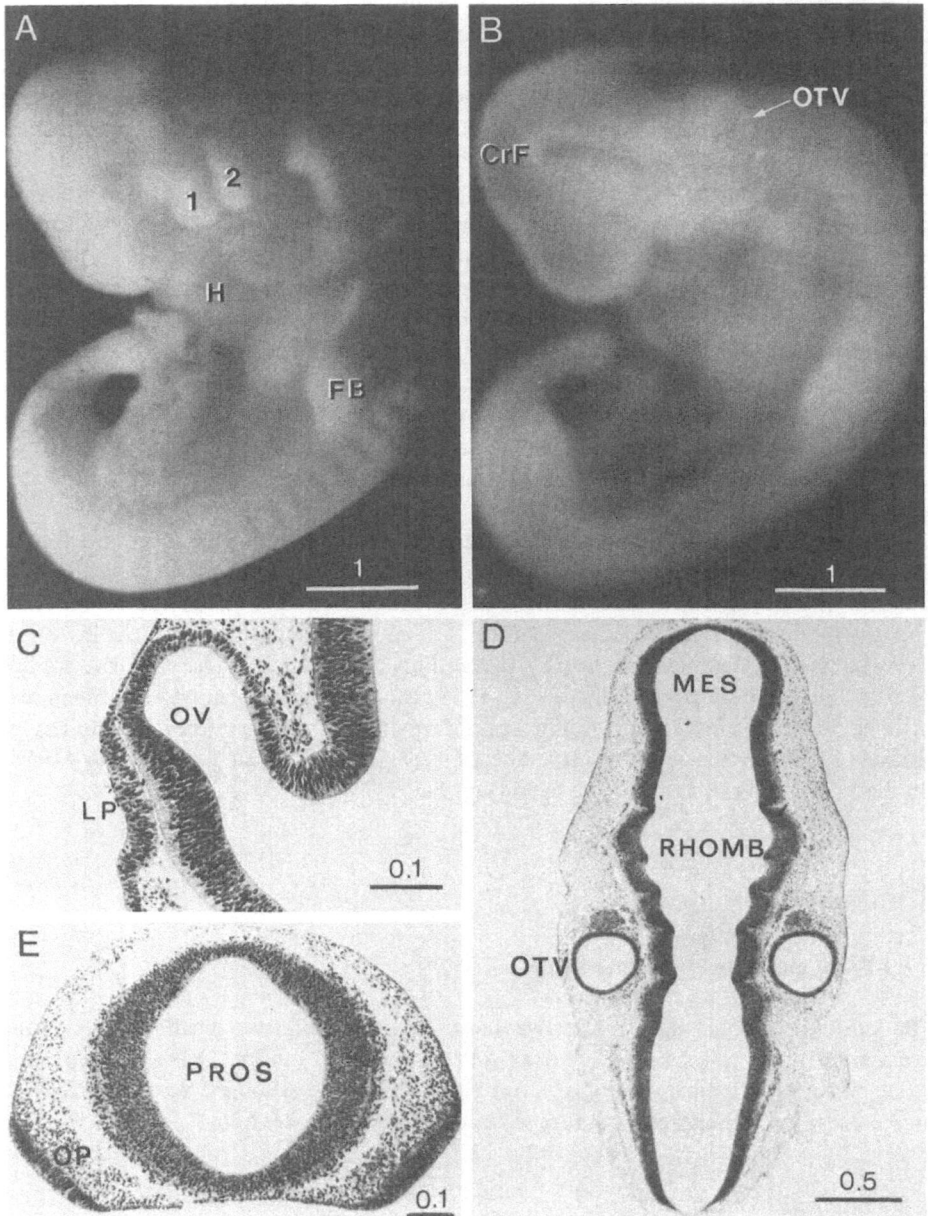

Fig. 1 A–E. Photomicrographs of stage 13 embryos, aged 32 ± 1 days p.c. A and B. Lateral views of the same embryo; in B cleared in methylbenzoate, C, D, and E. Frontal sections; in C of an older member of stage 13. Length of the *bars* noted in millimeters. Abbreviations: *CrF*, cranial flexure; *FB*, forelimb bud; *H*, heart bulge; *LP*, lens placode; *MES*, mesencephalon; *OP*, olfactory placode; *OTV*, otic vesicle; *OV*, optic vesicle; *PROS*, prosencephalon *RHOMB*, rhombencephalon; *1, 2* first and second branchial bars

be observed (Fig. 1A). A definite forelimb bud is present, whereas only a primordial hindlimb bud can be recognized. There is a short tail with a knoblike end, which is directed cranially.

3.1.3 Internal Features

In the embryos at stage 13 a lens placode is formed, consisting of 3–4 rows of cells, as a thickening of the single-cell layered epithelium of the surface ectoderm. In the older specimens a slight indentation of the lens placode can be observed (Fig. 1C). The laterally directed optic evagination shows a broad communication with the prosencephalon. A retinal disc is present, consisting of 6–7 rows of cells, which in the more advanced embryos bulges slightly into the lumen of the optic vesicle (Fig. 1C). The otic vesicle is completely closed and detached from the surface epithelium (Fig. 1D). A rudimentary endolymphatic appendage is present as a dorsomedial extension of the otocyst. A flat olfactory placode is present, consisting of 6–8 rows of cells (Fig. 1E). Its formation as a thickening of the surface ectoderm is similar to the process occurring both in the lens placode and the otic placode, which develops in developmental stage 10. Rathke's pouch, appearing in stage 12 as a small outgrowth of the stomodeum, starts to grow and is in close contact with the floor of the prosencephalon. In the CNS the three primary brain vesicles, the prosencephalon, mesencephalon, and rhombencephalon can be recognized (Fig. 1B). Although the prosencephalon shows the beginning of a dorsally directed outgrowth of the telencephalic area, the boundary between telencephalon and diencephalon cannot be defined. The form of the rostral part of the CNS is dominated by the deep flexura cranialis on top of which the mesencephalon is located (Fig. 1B). The rhombencephalon is characterized by its very thin roof. In microscopic sections the floor of the rhombencephalon shows a conspicuous metameric pattern (Fig. 1D) caused by the presence of prominent rhombomeres. The rather caudally situated flexura cervicalis marks the boundary with the spinal cord (Fig. 1B).

3.1.4 Comparison with Other Mammals

In Table 2 the main characteristics of rhesus monkey stage 13 embryos are compared with those in some other primates and rodents. It can be concluded that apart from the marmoset, the primates are very similar to one another both in age and in length at stage 13. The deviation of the marmoset was attributed by Phillips (1976) to a generally slower rate of development rather than to a delayed implantation (see also sect. 4). The latter suggestion was made by Butler (1972) for the galago, in which stage 13 appears to occur at the age of about 38 days. The two rodent species are a little less concordant than the primates: the mouse, represented by Theiler's (1972) stage 16 attains stage 13 2 days earlier than the Chinese hamster. From the brief studies on other mammalian species we have mentioned, it can tentatively be concluded that developmental stage 13 occurs in the guinea pig at about 16.5 days; in the rabbit at about 11 days, and in the rat at about 11.5–12 days. The length of the rat stage 13 embryos varies between 3.0 and 4.1 mm (Christie 1964).

In a comparison of the external features of the primate and rodent embryos two differences should be noted: the relatively earlier closure of the posterior neuropore

Table 2. Characteristics of mammalian stage 13 embryos

	Man	Baboon	Rhesus monkey	Marmo-set	Chinese hamster	Mouse
Age (days)	28–32	(28)[b] 29±1 (30)	(27) 28–30 (33)	61–67	$11^{1/2}$–12	10
Length (mm)	4.0–6.0	4.5–6.0	4.5–6.0	2.4–4.8	2.8–4.2	3.1–3.9
Posterior neuropore closed	+	+	+	+	–	–
Three branchial bars	+	+	+	+	+	+
Mandibular + maxillary processes	+	+	+	?	+	?c
Definite forelimb bud	+	+	+	+	+	+
Definite hindlimb bud	±a	±	±	±	+	+
Otic vesicle closed	+	+	+	+	+	+
Endolymphatic appendage	+	+	+	+	+	+
Onset lens indentation	–	–	±	–	+	+
Onset olfactory placode indentation		–	–	–	+	+
Outgrowth of Rathke's pouch	+	+	+	+	+	?
Three primary brain vesicles	+	+	+	+	+	+
Deep flexura cervicalis	+	+	+	+	+	+
Rhombomeres	+	+	+	?	?	?

a ±: present in older, absent in younger members.
b (n): minimum and maximum ages found.
c ?: not reported by the respective author.

in the primates and the somewhat earlier formation of the hindlimb bud in the rodents. As far as the internal features are concerned a striking resemblance can be noted between the primates and the rodents. Only the identation of both the lens and the olfactory placodes occurs later in primates than it does in rodents. In summary it can be stated that in the species analyzed the embryos representative of developmental stage 13 are highly comparable.

3.2 Stage 14

3.2.1 Postconceptional Age and Length

Developmental stage 14 is represented by five rhesus monkey embryos, ranging in age from 30 ± 1 to 32 ± 1 days p.c. One embryo aged 28 ± 1 days p.c. had also to be included in this stage. The greatest length of the specimens varied from 6.0 to 8.0 mm.

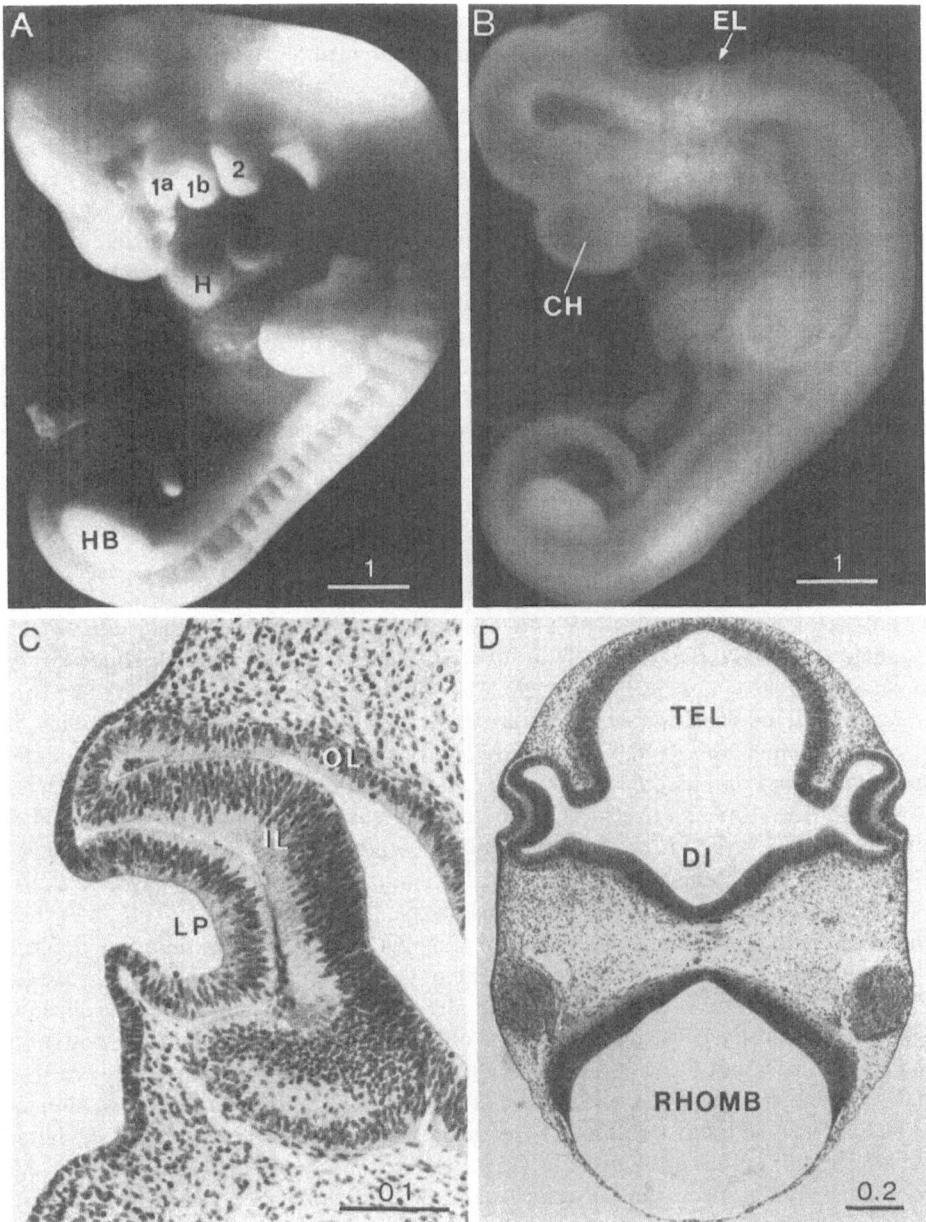

Fig. 2 A–D. Photomicrographs of stage 14 embryos, aged 30 ± 1 days p.c. A and B. Lateral views of the same embryo; in B cleared in methylbenzoate, C. Frontal section; D. Horizontal section. Length of the *bars* noted in millimeters. Abbreviations: *CH*, cerebral hemisphere; *DI*, diencephalon; *EL*, endolymphatic appendage; *H*, heart bulge; *HB*, hindlimb bud; *IL*, inner layer of optic cup; *LP*, lens placode; *OL*, outer layer of optic cup; *RHOMB*, rhombencephalon; *TEL*, telencephalon; *1a*, maxillary process; *1b*, mandibular process; *2*, hyoid process

3.2.2 External Features

The outer form of embryos belonging to developmental stage 14 is characterized by its angular shape (Fig. 2A). In addition to the cervical and sacral flexures a slight thoracic curvature is present. A striking feature of this stage is the straight trunk in between the thoracic and sacral flexures. In addition, the somites are very manifest structures characterizing the outside image of stage 14 embryos. The frontal part of the head is in contact with the heart bulge or even extends lateral to it. The various subdivisions of the heart can be discerned from the outside. The branchial bars are very prominent, particularly the processus maxillaris, which has extended compared with stage 13. Both a deep nasal pit and a lens pit are visible on the outside (Fig 2A). The superior contour of the head is determined by the CNS lying directly underneath the skin (e.g. Fig. 2A and B). The forelimb bud has enlarged considerably and shows the first sign of a separation into a proximal part and a distal part. A distinct hindlimb bud is present. The tail has lengthened and now curves around the hindlimb bud, its tip pointing dorsally (Fig. 2A and B).

3.2.3 Internal Features

A characteristic feature of stage 14 is the deep indentation of the lens placode (Fig. 2C). The optic cup is formed by the progressive invagination of the retinal disc, which parallels the indentation of the lens. Ventrally, the invagination of the optic cup is incomplete, leaving a wide optic fissure. The width of the inner layer of the optic cup, which represents the former retinal disc and the future neural layer of the retina, is about three times the width of the outer layer of the optic cup (Fig. 2C). The inner layer consists of an inner zone containing 5–6 rows of nuclei and an outer zone free of nuclei. The outer layer of the optic cup, the future pigment layer of the retina, consists of 2–3 layers of cells arranged into a pseudostratified epithelium. The cavity of the optic cup communicates with the lumen of the forebrain by way of a short, wide optic stalk (Fig. 2D). The otic vesicle has enlarged considerably (Fig. 3A). The endolymphatic appendage has lengthened (Fig. 3A and B), now forming the definite endolymphatic duct that is even recognizable in the embryo cleared in methylbenzoate (Fig. 2B). The thickness of the walls of the otic vesicle and the endolymphatic appendage varies regionally (Fig. 3A and B). The olfactory placode shows a deep indentation and consists of about 8 rows of cells (Fig. 3C). Rathke's pouch is considerably enlarged and its connection with the stomodeum is narrowing (Fig. 3D). At the site of contact of Rathke's pouch with the floor of the diencephalon a slight depression of the latter marks the primordial infundibulum (Fig. 3D). The morphology of the CNS is still dominated by the deep flexura cranialis and the distinct flexura cervicalis (Fig. 2B). A slight flexura pontina has also developed. In the prosencephalon the dorsal extension of the telencephalic area (which started in the preceding stage) is now attended by a bilateral evagination (Fig. 2B). Hence a subdivision into telencephalon and diencephalon is effected. Yet, both dorsally and frontally the telencephalon still consists of a single bulge (Fig. 2D) whereas laterally and caudally two hemisphere bulges can be observed (Fig. 2B). At the junction between the mesencephalon and the rhombencephalon the narrowing isthmus rhombencephali becomes apparent (Fig. 2B). In the rhombencephalon the rhombomeres are still present, while the cerebellar plates have started to develop (Fig. 2B).

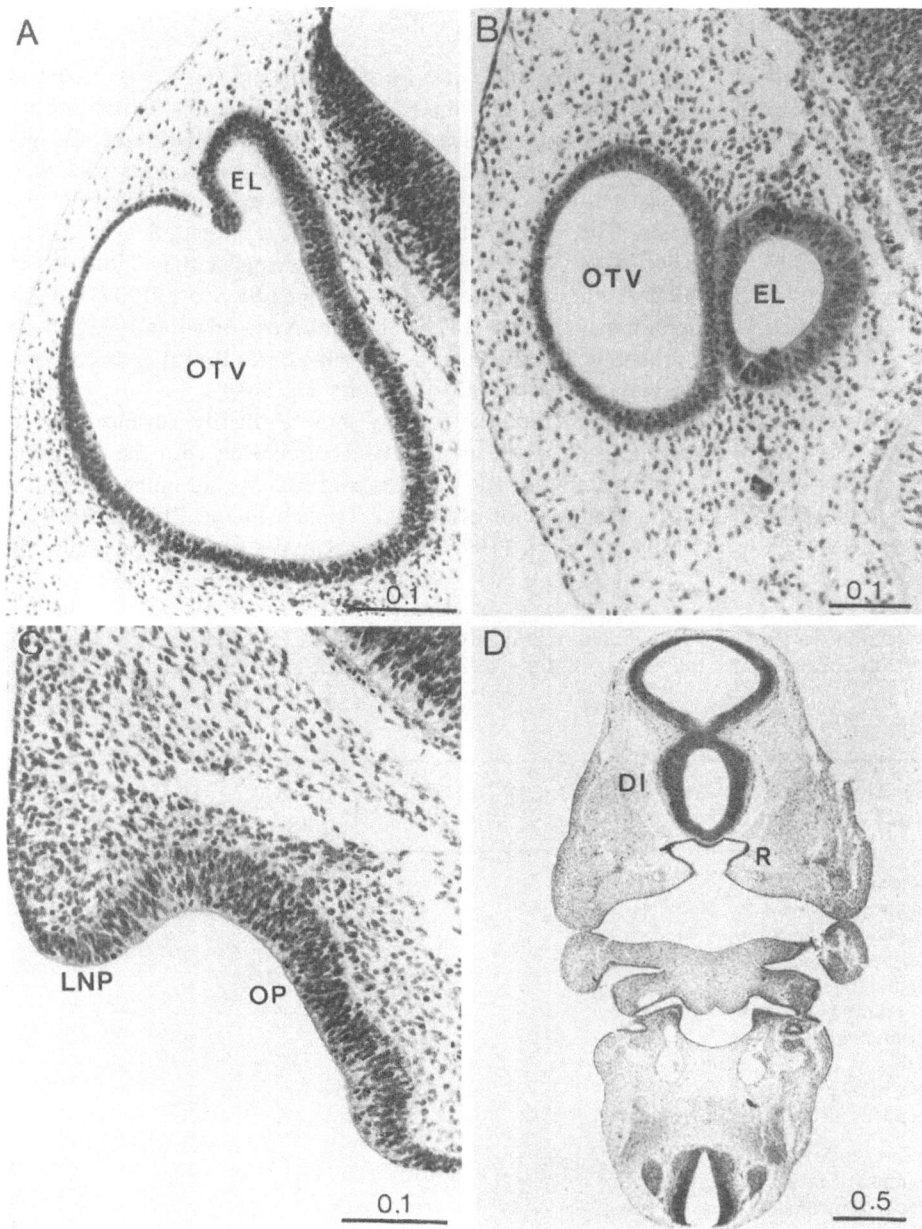

Fig. 3 A–D. Photomicrographs of stage 14 embryos. A. Horizontal section of an embryo aged 32 ± 1 days p.c.; B, C and D. Frontal sections of an embryo aged 30 ± 1 days p.c. Length of the *bars* noted in millimeters. Abbreviations: *DI*, diencephalon; *EL*, endolymphatic appendage; *LNP*, lateral nasal process; *OP*, olfactory placode; *OTV*, otic vesicle; *R*, Rathke's pouch

3.2.4 Comparison with Other Mammals

The main characteristics of developmental stage 14 of the mammalian species analyzed are summarized in Table 3. Among the primates a striking congruity exists in both the age and the length of the representative embryos. Only the marmoset diverges considerably in being much older and smaller. Similar observations were presented by Butler (1972) on the galago, in which stage 14 embryos are about 38 days old and have a length of 2—5 mm. The mouse stage 14 embryos, represented by Theiler's (1972) stage 17, are still 2 days younger as well as a little smaller than those of the Chinese hamster. The 17.5-day guinea pig embryos described by Scott (1937) exhibit the characteristics of developmental stage 14. Rabbit embryos (Edwards 1968) about 11—12 days of age are probably also at this stage; the same holds for rat embryos aged 12—13 days and 4.1—5.8 mm long, as described by Christie (1964).

Regarding external characteristics, the rodents show a slightly advanced development of the forelimb bud and of the nasal pit in comparison with the primates. Concerning the internal characteristics the primates and rodents are much the same. The only striking feature is the closure of Rathke's pouch in the Chinese hamster as described by ten Donkelaar et al. (1979), whereas in the mouse an opening of

Table 3. Characteristics of mammalian stage 14 embryos

	Man	Baboon	Rhesus monkey	Marmoset	Chinese hamster	Mouse
Age (days)	31—35	(27) 30±1 (32)	(27) 30—32 (33)	63—?	12—$12^{1/2}$	$10^{1/2}$
Length (mm)	5—7	6—7	6—8	4.1—?	4.6—5.1	3.5—4.9
Posterior neuropore closed	+	+	+	+	+	+
Mandibular + maxillary processes	+	+	+	+	+	+
Subdivision forelimb bud	±	±	±	±	+	+
Definite hindlimb bud	+	+	+	+	+	+
Deep nasal pit	±	±	+	−	+	+
Deep lens pit	+	+	+	±	+	+
Definite endolymphatic duct	+	+	+	+	+	+
Deep lens indentation	+	+	+	±	+	+
Deep indentation olfactory placode	±	±	±	±	+	±
Closure Rathke's pouch	−	−	−	−	+	−
Evagination telencephalic hemispheres	+	+	+	+	+	+
Flexura pontina	±	±	+	?	+	+
Rhombomeres	+	+	+	?	?	+

14

Rathke's pouch is still present during stage 16, as reported by Theiler (1972). In the primates the closure of Rathke's pouch occurs even later, namely during stage 18, as will be described in Sect. 3.6.4. On the other hand, a very characteristic feature present in the embryos of all species described is the deep indentation of the lens placode (see also Pei and Rhodin 1970; Jackson 1976).

3.3 Stage 15

3.3.1 Postconceptional Age and Length

The postconceptional age of developmental stage 15 deduced from the rhesus monkey embryos examined varied between 30 ± 1 and 33 ± 1 days p.c. The maximum age recorded was represented by a specimen aged 34 ± 1 days p.c. The greatest length of the embryos measured ranged from 7 to 9 mm.

3.2.2 External Features

The outer form of the embryos at stage 15 is characterized by a C-shaped body axis still dominated by the cervical flexure (Fig. 4A). Compared with the preceding stage the specimens have a much more rounded appearance; the somites are prominent structures from the cervical to the coccygeal region (Fig. 4A and B). The mandibular and hyoid processes remain prominent, the latter showing tentative auricular hillocks (Fig. 4A). The maxillary process has considerably enlarged but is separated from the lateral nasal process by a deep nasolacrimal cleft. The deep, oval olfactory pit, which is located rather laterally, is bordered by the protruding medial and lateral nasal processes. The various subdivisions of the brain can be recognized exteriorly (e.g., Fig. 4A and B). The forelimb bud is subdivided into a proximal arm segment and a distal hand segment, the latter directed more medially (Fig. 4A). The hindlimb bud has increased in length but no subdivision can be observed. The tail has lengthened compared with the preceding stage and in some advanced specimens shows a knob at the end.

3.3.3 Internal Features

The most conspicuous internal feature of developmental stage 15 is the closure of the lens vesicle (Fig. 4C). In some of the specimens a narrow pore is still present, through which the lumen of the lens communicates freely with the exterior (Fig. 4D). In the more advanced embryos the lens vesicle is closed, but it is still attached to the overlying surface ectoderm (Fig. 4C). The cells of the lens vesicle are arranged radially; those of the medial wall are elongated compared with those of the lateral wall. The optic cup is still attached to the diencephalon by way of a short, wide optic stalk (Fig. 4D). The retinal fissure is rather wide, its mesenchymal contents being continuous with those of the future lentoretinal space. In microscopical sections of embryos of this stage the first pigment granules can be observed within the outer pigment layer of the retina. The pigment is also evident in the embryos cleared in methylbenzoate

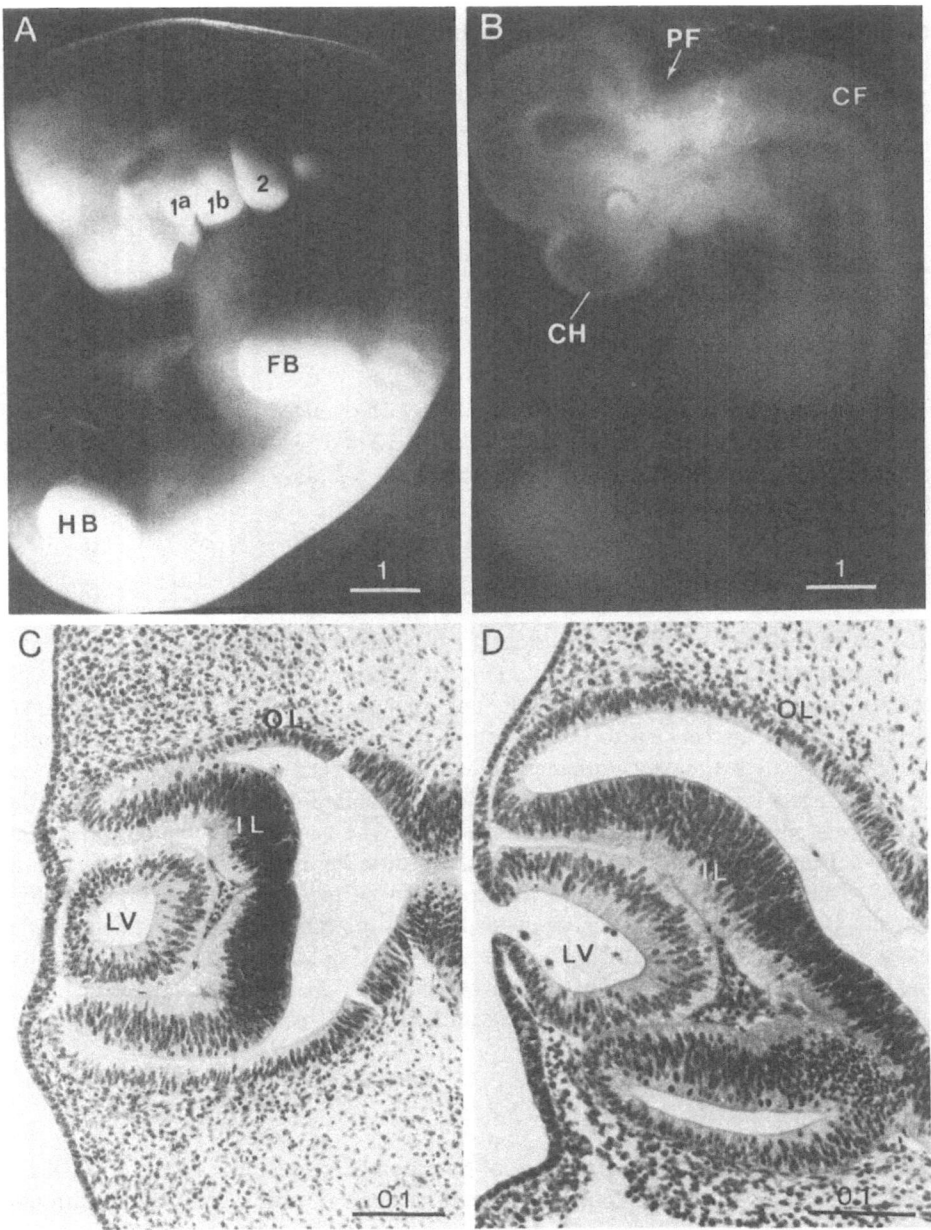

Fig. 4 A–D. Photomicrographs of stage 15 embryos, aged 30 ± 1 days p.c. A and B. Lateral views of two embryos; B. Cleared in methylbenzoate, C. Horizontal section of embryo A, D. Frontal section of embryo B. Length of the *bars* noted in millimeters. Abbreviations: *CF*, cervical flexure; *CH*, cerebral hemisphere; *FB*, forelimb bud; *HB*, hindlimb bud; *IL*, inner layer of optic cup; *LV*, lens vesicle; *OL*, outer layer of optic cup; *PF*, pontine flexure; *1a*, maxillary process; *1b*, mandibular process; *2*, hyoid process

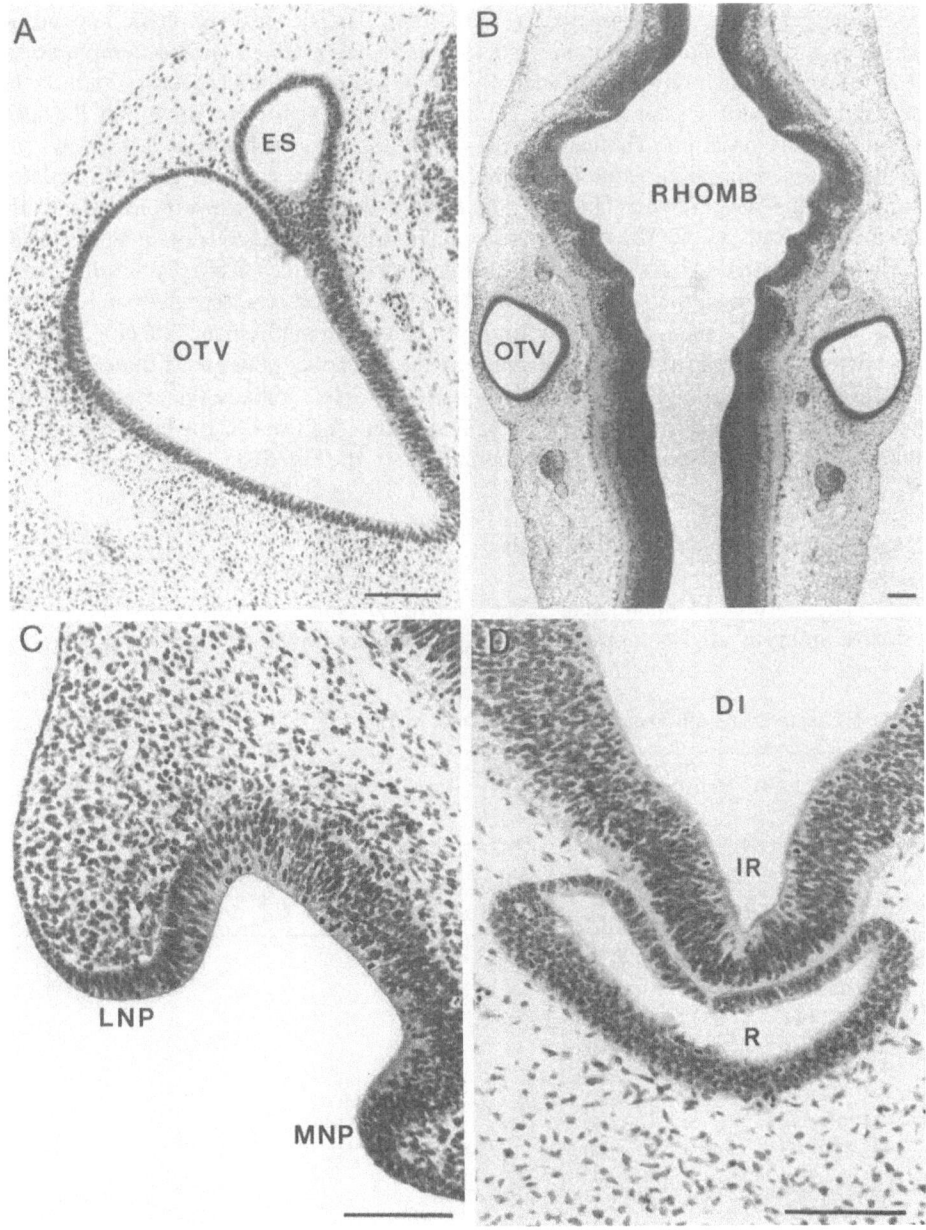

Fig. 5 A–D. Photomicrographs of stage 15 embryos, aged 30 ± 1 days p.c. A. Horizontal section, B, C and D. Frontal sections. Length of the *bars* 0.1 mm. Abbreviations: *DI*, diencephalon; *ES*, endolymphatic sac; *IR*, infundibular recess; *LNP*, lateral nasal process; *MNP*; medial nasal process; *OTV*, otic vesicle; *R*, Rathke's pouch; *RHOMB*, rhombencephalon

(Fig. 4B). The otocyst has attained a more triangular shape in which the cochlear and vestibular regions become apparent (Fig. 5A and B). The endolymphatic appendage, which is a diverticulum of the vestibular region, has attained an endolymphatic sac (Fig. 5A). The indentation of the nasal placode has proceeded especially at the edge of the medial nasal process (Fig. 5C). Numerous mitotic figures are present at the outer surface of the nasal pit. The caudal part of Rathke's pouch has spread laterally, surrounding the definite recessus infundibuli (Fig. 5D). The shape of the CNS is determined by the deep flexura cranialis, the slightly deepening flexura pontina, and the flexura cervicalis (Fig. 4B). The telencephalic hemispheres are dorsolateral bulges, dorsally separated by the sulcus interhaemisphaericus, which frontally is still absent. The diencephalon shows some successive bulges or neuromeres, especially in its caudal part, which can be recognized in the embryos cleared in methylbenzoate (Fig. 4B) and in sagittal sections. In the rhombencephalon the cerebellar plates have thickened (Fig. 5A) and cover the lateral recesses of the fourth ventricle. Caudally the cerebellar plates border upon the thin roof of the rhombencephalon. The floor of the latter part of the brain is still characterized by the rhombomeric pattern (Fig. 5B).

3.3.4 Comparison with Other Mammals

The main features of mammalian stage 15 embryos are summarized in Table 4. The primate embryos at developmental stage 15 vary more or less in age: on average the

Table 4. Characteristics of mammalian stage 15 embryos

	Man	Baboon	Rhesus monkey	Marmo-set	Chinese hamster	Mouse
Age (days)	35−38	(30) 31±1 (32)	(29) 30−33 (35)	61−73	$12^{1/2}$−13	11
Length (mm)	7−9	6−8	7−9	6.8−7.8	5.1−6	5−6
Subdivision fore-limb bud	+	+	+	+	+	+
Subdivision hind-limb bud	−	−	−	−	+	±
Deep nasal pit	+	+	+	+	+	+
Lens pit closed	+	+	+	+	+	+
First retinal pig-ment (MB)	+	?	+	−	+	?
Cochlear + vesti-bular regions	+	+	+	+	+	+
Closure lens vesicle	+	+	+	+	+	+
First retinal pig-ment	+	+	+	−	+	+
Deep indentation olfactory placode	+	+	+	+	+	+
Rathke's pouch closed	−	−	−	−	+	−
Flexura pontina	+	+	+	?	+	+
Rhombomeres	+	+	+	?	?	?

human embryos are a little older than those of the baboon and the rhesus monkey, whereas the marmoset is much older. Nevertheless, all primate embryos exhibit about the same length; compared with the preceding stage the marmoset has caught up with the other primates. No data are available on the galago. In the Chinese hamster and the mouse (Theiler's stage 18) the ages reported were still different but the same lenghts were recorded for stage 15 embryos. In the guinea pig stage 15 embryos are about 18 1/2 days old (Scott 1937), whereas in the rabbit comparable embryos are probably aged about 12–13 days. In the rat, stage 15 embryos are aged 13–13 3/4 days, varying in length between 5.8 and 7.9 mm, as reported by Christie (1964).

The external characteristics of the stage 15 embryos in all species analyzed concur fairly well. Only the subdivision of the hindlimb bud that is recognizable in the rodents is still absent in the primates. Similar results were obtained for the forelimb bud as described in the preceding stage. The internal characteristics of the embryos in the various species correspond even better. Except for the presence of a closed Rathke's pouch in the Chinese hamster, as already discussed in the description of stage 14, all internal features analyzed are identical. The most useful distinguishing internal characteristics of embryos representative of developmental stage 15 are the closure of the lens vesicle and the presence of the first retinal pigment.

3.4 Stage 16

3.4.1 Postconceptional Age and Length

The three rhesus monkey embryos examined as representatives of developmental stage 16 had an age of 32 ± 1 to 34 ± 1 days p.c. The oldest specimen found to be at stage 16 was aged 36 ± 1 days. The greatest length of the stage 16 embryos varied between 7 and 11 mm.

3.4.2 External Features

The external form of the embryos is still dominated by the cervical flexure that now approximates a right angle (Fig. 6A and B). The heart-liver bulge is relatively enlarged compared with that in the preceding stage 15 (e.g., Fig. 4A and B). Except for the cervical part of the body the somites are recognizable up to the tip of the tail (Fig. 6A and B). The hyoid bar is characterized by the presence of distinct auricular hillocks. The maxillary process extends toward the lateral nasal process forming a definite nasolacrimal groove. The nasal pits bordered on by the medial and lateral nasal processes have a more medial position than in the preceding stage. The presence of retinal pigment is faintly recognizable on the outside (Fig. 6A) but clearly visible in the pictures of the embryos cleared in methylbenzoate (Fig. 6B). In these pictures the open circle of pigmentation, caused by the arrangement of the pigment granules in the retina (see Sect. 3.4.3), is characteristic for stage 16 in contradistinction to the quadrangular arrangement in stage 17 embryos (e.g., Figs. 6B and 8B). The distal part of the forelimb bud is transformed definitively into a round handplate (Fig. 6A), whereas the hindlimb bud has extended and in the older specimens shows the first sign of a subdivision. The tail has lengthened and in some embryos at this stage exhibits a knoblike end.

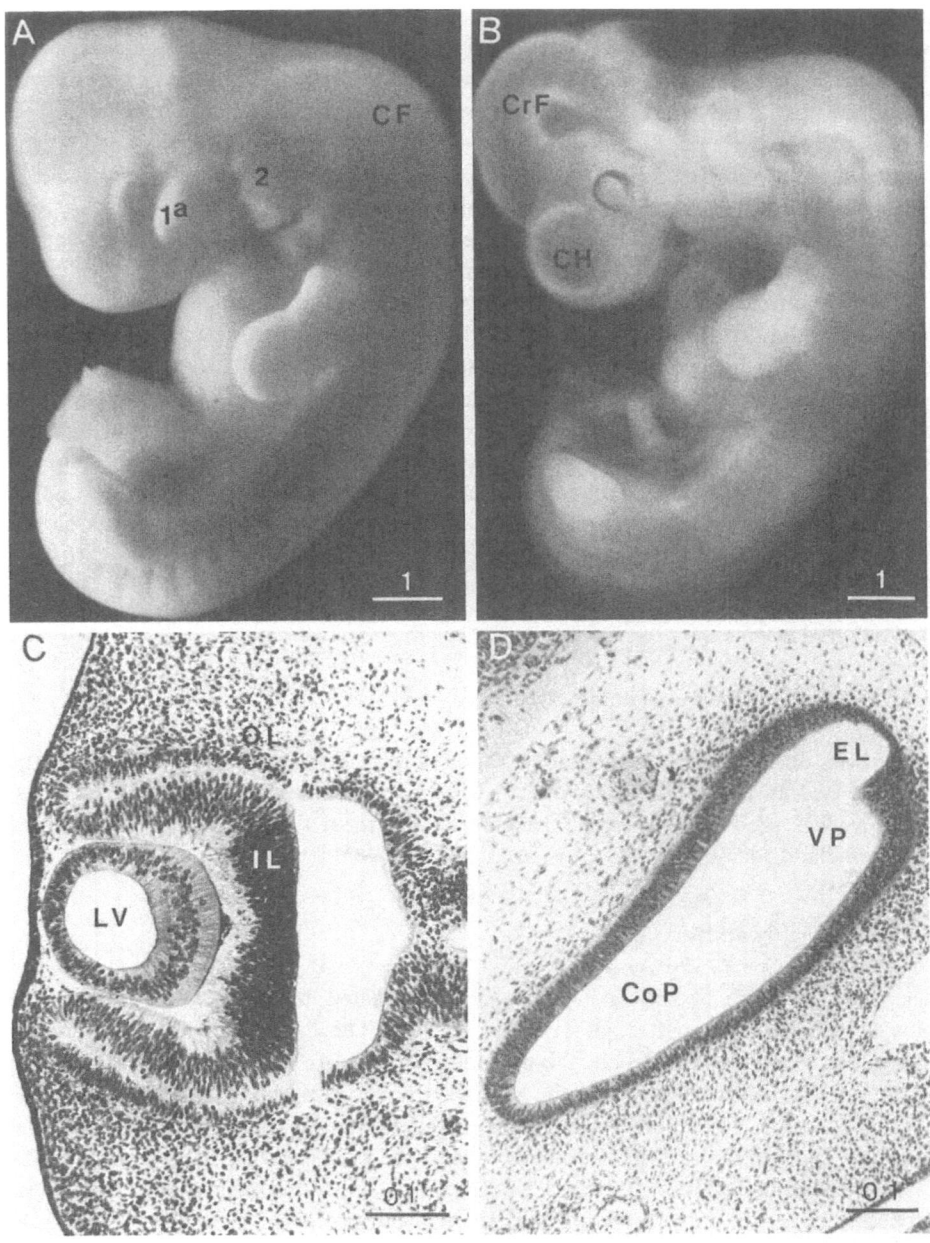

Fig. 6 A–D. Photomicrographs of stage 16 embryos, aged 32 ± 1 days p.c. A and B. Lateral views of the same embryo; in B cleared in methylbenzoate, C and D. Horizontal sections. Length of the *bars* noted in millimeters. Abbreviations: *CF*, cervical flexure; *CH*, cerebral hemisphere; *CoP*, cochlear pouch; *CrF*, cranial flexure; *EL*, endolymphatic duct; *IL*, inner layer of optic cup; *LV*, lens vesicle; *OL*, outer layer of optic cup; *VP*, vestibular pouch; *1a*, maxillary process; *2*, hyoid process

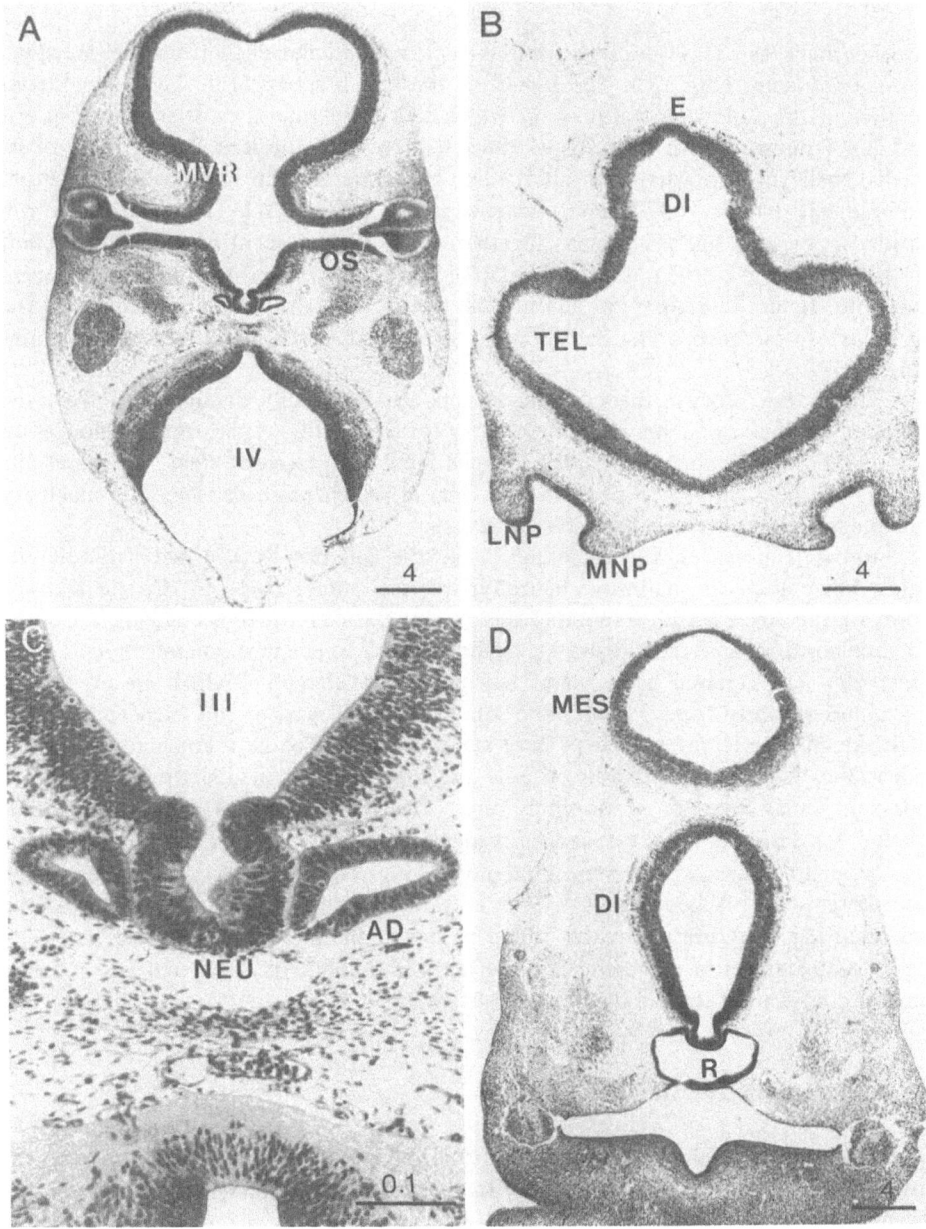

Fig. 7 A–D. Photomicrographs of stage 16 embryos, aged 32 ± 1 days p.c. A and C. Horizontal sections, B and D. Frontal sections. Length of the *bars* noted in millimeters. Abbreviations: *AD*, adenohypophysis; *DI*, diencephalon; *E*, epiphysis; *LNP*, lateral nasal process; *MES*, mesencephalon; *MNP*, medial nasal process; *MVR*, medial ventricular ridge; *NEU*, neurohypophysis; *OS*, optic stalk; *R*, Rathke's pouch; *TEL*, telencephalon; *III*, third ventricle; *IV*, fourth ventricle

21

3.4.3 Internal Features

At this stage the lens vesicle is invariably closed and detached from the overlying surface ectoderm (Fig. 6C). The lumen of the lens is spherical in shape, in contrast to that in the following stage 17, in which it is crescentic in cross section (Fig. 9A and B). Compared with the cells of the lateral wall of the lens vesicle, those of its medial wall are somewhat elongated, their nuclei holding an intermediate position. The elongated optic stalk shows a narrowing lumen (Fig. 7A). The optic fissure is a slitlike groove, its lips lying close together. Numerous pigment granules are present in the outer layer of the optic cup (Fig. 6C); however, their number diminishes toward the optic fissure and they are absent near the optic stalk. This arrangement of the granules causes the opening in the circle of pigment in the methylbenzoate pictures (Fig. 6B).

Within the otocyst the cochlear pouch can be readily distinguished from the rounded vestibular pouch by its more oblong form (Fig. 6D). Apart from its elongation the endolymphatic duct has hardly changed from the preceding stage. The nasal pits are transformed into deep pockets (Fig. 7B) in which more caudally the maxillary and medial nasal processes have started to fuse.

Rathke's pouch has considerably enlarged and has formed two frontolateral lobes bilateral to the definitive neurohypophyseal anlage (Fig. 7A, C, and D). The form of the CNS is still dominated in turn by the deep flexura cranialis, the slight flexura pontina, and the flexura cervicalis, which forms approximately a right angle (Fig. 6B). The cerebral hemispheres are distinct vesicles now, which are even recognizable exteriorly (e.g., Fig. 6A and B). In the basal part of the telencephalic wall a thickening is present, which in later stages turns out to be the medial ventricular ridge (Fig. 7A); in the literature this is also known as the medial striatal ridge. The interventricular foramen of Monro is very wide (Fig. 7A and B). Within the diencephalon a primordial epiphysis can be recognized in its roof (Fig. 7B), while its floor has evaginated forming the neurohypophysis (Fig. 7A, C, and D). The lumen of the neurohypophysis is continuous with the third ventricle. Decussating fibers of the trochlear nerve are present at the caudal border of the mesencephalic roof. The cerebellar plates are clearly present (Fig. 6B); the rhombomeres, although still present, are less conspicuous than in the preceding stage 15.

3.4.4 Comparison with Other Mammals

In Table 5 in which Theiler's mouse stage 19 is presented, the main features of mammalian stage 16 embryos are summarized. In the primates the human stage 16 embryos are approximately 6 days older than the baboon and rhesus monkey embryos, whereas the marmoset embryos are even twice as old. No data are available on galago stage 16 embryos. Regarding the rodents, Chinese hamster embryos still attain developmental stage 16 2 days later than mouse embryos, although on average they are smaller than the latter. From the available data on the other species it seems reasonable to assume the following ages for stage 16 embryos: for the guinea pig (Scott 1937) about 19 1/2 days; for the rabbit (Edwards 1968) about 13–14 days; for the rat (Christie 1964) 13 3/4–14 days. The length of these rat embryos varies between 7.9 and 9.4 mm.

Table 5. Characteristics of mammalian stage 16 embryos

	Man	Baboon	Rhesus monkey	Marmo-set	Chinese hamster	Mouse
Age (days)	37−42	(32) 33±1 (34)	(31) 32−34 (37)	66−83	$13^{1/2}$	$11^{1/2}$
Length (mm)	8−11	7−9	7−11	8.3−9.2	5.6−6.3	6−7
Cervical flexure approx. 90°	+	+	+	?	+	+
Somites recognizable	+	+	+	+	+	+
Auricular hillocks on hyoid bar	+	+	+	+	+	+
Definite naso-lacrimal groove	+	+	+	+	+	+
Retinal pigment visible	±	±	±	±	±	±
Definite round handplate	+	+	+	+	+	+
Subdivision hind-limb bud	+	±	±	±	+	±
Lens vesicle closed, detached from ectoderm	+	+	+	+	+	+
Medial wall of lens starts to thicken	+	+	+	+	+	+
Optic stalk elongated	+	+	+	?	+	+
Optic fissure slit-like	+	+	+	?	?	+
Definite retinal pigment	+	+	+	+	+	+
Separation cochlear and vestibular pouches	+	+	+	+	?	+
Nasal pits form deep pockets	+	+	+	+	+	+
Medial and lateral nasal processes start fusion	+	+	+	+	?	+
Rathke's pouch closed	−	−	−	−	+	−
Rathke's pouch shows two frontal lobes	+	+	+	?	+	?
Neurohypophysis starts evagination	+	+	+	+	+	+
Cerebral hemispheres are distinct vesicles	+	+	+	+	+	+
Cerebellar plates present	+	+	+	+	+	+

As regards the external features summarized in Table 5, it can be noted that hardly any difference exists between the mammalian species analyzed. In comparison with stages 15 and 17 the main discriminative external characteristics of stage 16 embryos are the presence of auricular hillocks on the hyoid bar and of a definite round handplate.

The great conformity of mammalian stage 16 embryos is even accentuated by the great number of internal characteristics, as listed in Table 5, which are present in all species analyzed. The most characteristic internal features of stage 16 embryos are: the presence of a closed lens vesicle which is detached from the surface ectoderm, the presence of definitive pigment granules in the outer pigment layer of the retina, and the started evagination of the neurohypophysis from the floor of the diencephalon.

3.5 Stage 17

3.5.1 Postconceptional Age and Length

In total six rhesus monkey embryos at developmental stage 17 were examined. Their postconceptional age varied between 34 ± 1 days and 36 ± 1 days. The minimum age of this stage is represented by an embryo aged 33 ± 1 days, the maximum age being found in a specimen aged 37 ± 1 days. The greatest length of the stage 17 embryos ranged from 9 to 12 mm.

3.5.2 External Features

The external form of the embryos at stage 17 has hardly changed compared with that in the preceding stage 16 (e.g., Figs. 6A and 8A). The cervical flexure has become more acute; the somites, although still recognizable, are less conspicuous structures than in stage 16. The maxillary process extends up to the lateral nasal process, and the nasomaxillary and nasolacrimal grooves are formed. The deep nasal pits open more medially, the primitive nostrils being hardly visible in profile views. Apart from those on the hyoid bar, auricular hillocks are now also present on the mandibular bar (Fig. 8A). The mandibular and hyoid bars enclose the slitlike auditory groove.

Within the handplate a subdivision can be observed into a central convex carpal part and a peripheral flat fingerplate. In the older representatives of stage 17 the first sign of fingerrays is visible in the latter peripheral part of the handplate. A striking tubercle marks the boundary between the forearm and upperarm segments of the forelimb (Fig. 8A and B). The hindlimb shows an annular constriction separating the elongated leg segment from the foot segment. In the older members of the group the latter part is more flattened thereby having the character of a real footplate. The length of the tail has increased considerably; its tip may even reach up to the level of the forehead.

3.5.3 Internal Features

The most outstanding internal characteristic of stage 17 embryos is the crescentic shape of the lens cavity (Fig. 9A and B). The cells of the inner wall of the lens, form-

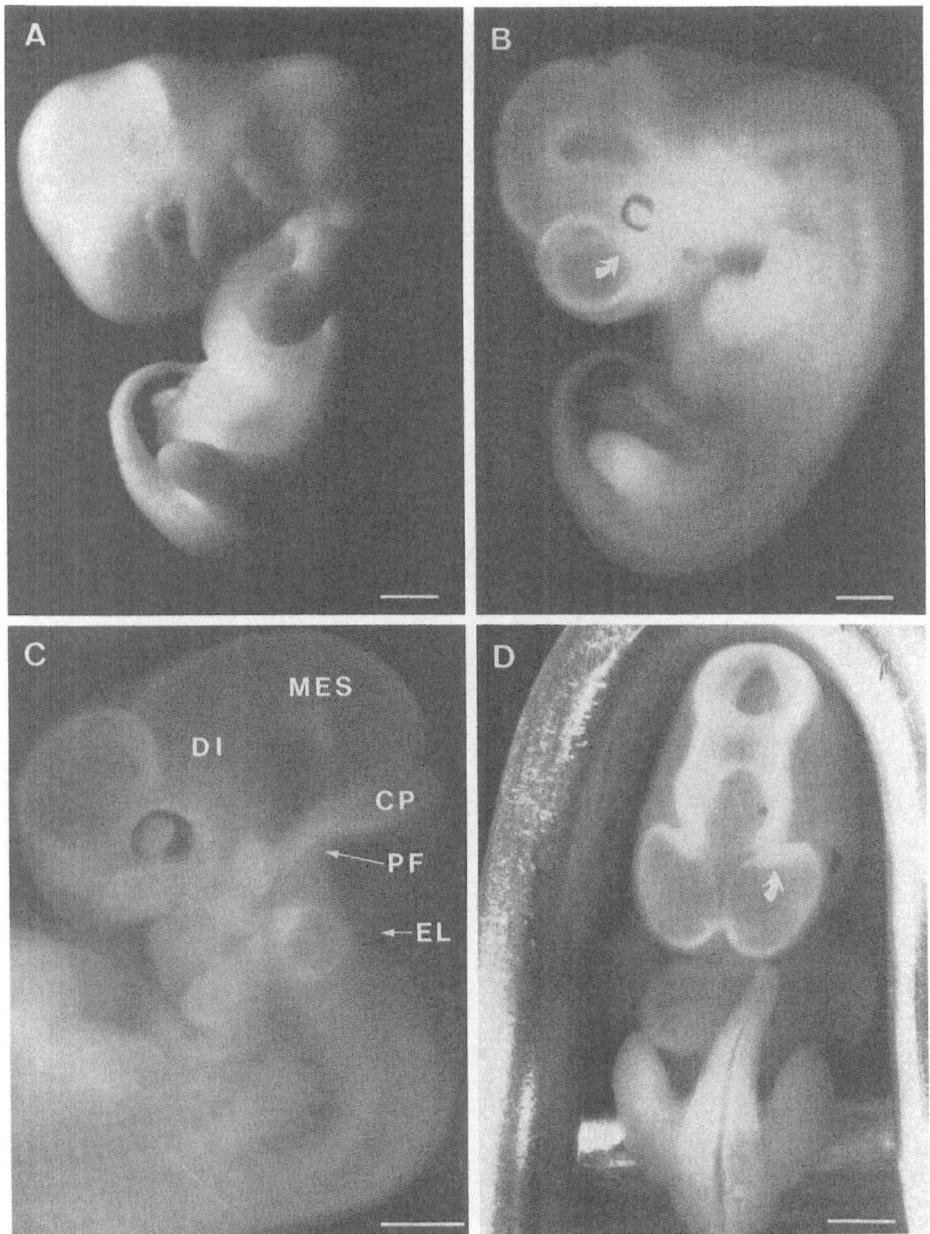

Fig. 8 A–D. Photomicrographs of stage 17 embryos. **A** and **B**. Lateral views of an embryo aged 33 ± 1 days p.c.; C. Lateral view of an embryo aged 35 ± 1 days p.c., D. Ventral view of an embryo aged 36 ± 1 days p.c., in B, C and D cleared in methylbenzoate. *Arrows* indicate thickened basolateral parts of the telencephalic hemispheres. Length of the *bars* 1 mm. Abbreviations: *CP*, cerebellar plate; *DI*, diencephalon; *EL*, endolymphatic duct; *MES*, mesencephalon; *PF*, pontine flexure

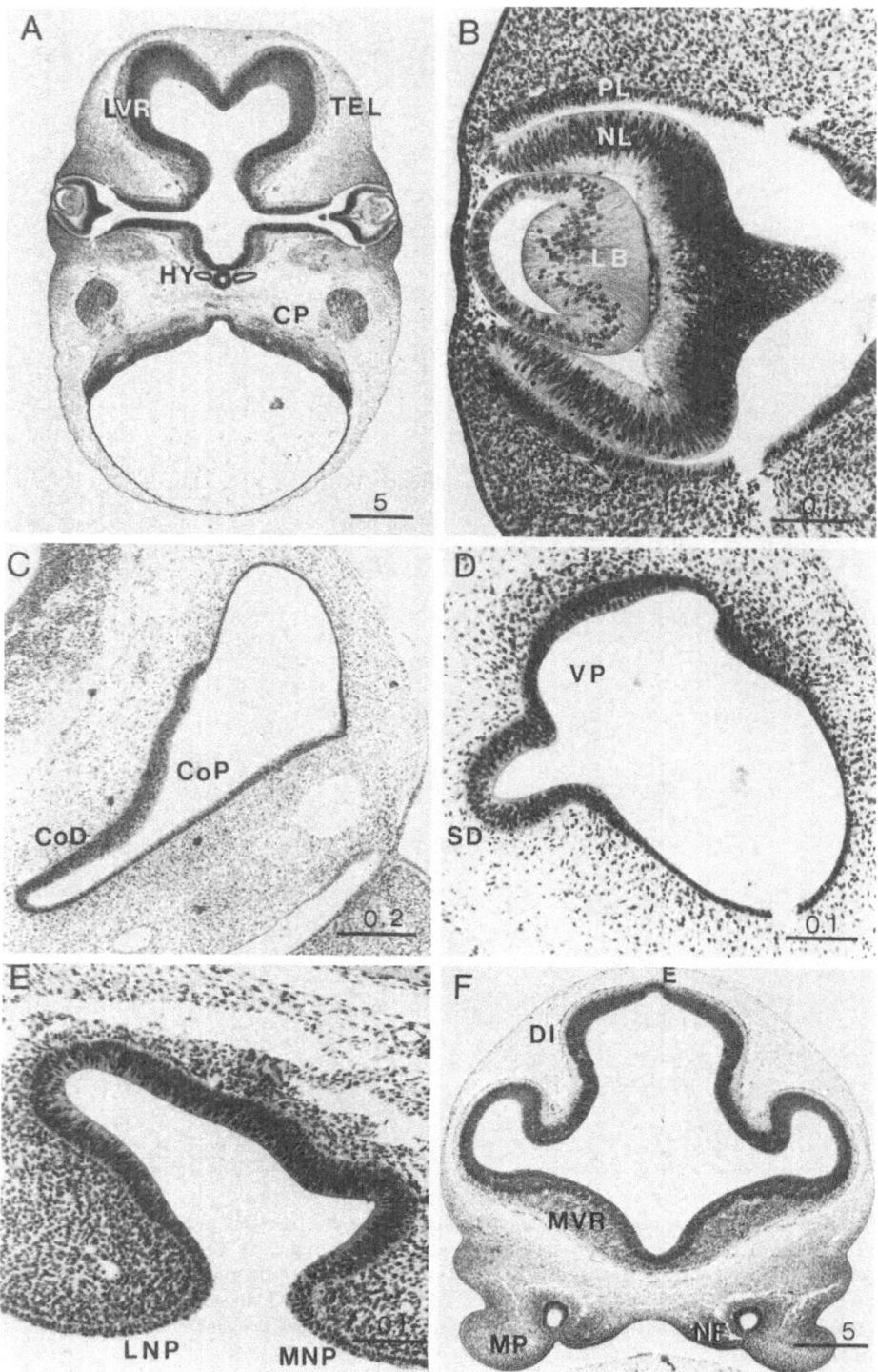

Fig. 9A–F

ing the future lens body, have lenghtened and protrude into the lumen of the lens representing primitive lens fibers. In the optic cup the lips of the optic fissure have started to fuse; the optic stalk exhibits a lumen over its full length (Fig. 9A). The number of retinal pigment granules has considerably increased compared with the preceding stage, but they are still absent near the optic stalk. This arrangement accounts for the opening in the pigmented ring, which at this stage is quadrangular, as can be seen in the pictures of the embryos cleared in methylbenzoate (Fig. 8B and C). The endolymphatic duct has lengthened and is also easily recognizable in such pictures (Fig. 8C). The cochlear pouch has extended forming the primordial cochlear duct (Fig. 9C), and the primordia of the semicircular ducts are visible as outpouchings of the vestibular pouch (Fig. 9D).

The olfactory pits have extended laterally (Fig. 9E) and caudally where they end blindly. The epithelia of the maxillary and the medial nasal process are fused to form a longitudinal septum, called the nasal fin (Fig. 9F). Caudally the epithelium of the fused maxillary and the medial nasal processes is continuous with that lining the roof of the primitive oral cavity. The two rostral evaginations of the adenohypophysis on either side of the neurohypophysis have extended (Fig. 9A). The neurohypophysis is now a rostral fingerlike projection of the floor of the diencephalon, its lumen being continuous with the third ventricle.

Because of its proportional growth the overall appearance of the CNS has hardly changed compared with that of the preceding stage (e.g., Figs. 8B, C, and 6B). Nevertheless, a deepening of the flexura pontina and an expansion of the telencephalic hemispheres can be recognized. The latter not only proceeds in a dorsal but also in frontal and caudal directions, as is clearly illustrated in Fig. 8D. The thickened basolateral parts of the hemispheres are clearly visible in the pictures of the embryos cleared in methylbenzoate (Fig. 8B, C, and D). In the sections of these parts of the telencephalon the first sign of a second ventricular ridge can be recognized lateral to the ventricular ridge already present in the previous stage (Fig. 9A). From now on these ridges will be called the medial ventricular ridge and the lateral ventricular ridge. The epiphysis is a small protrusion of the diencephalic roof (Fig. 9F); caudal to it, the first fibers of the commissura posterior are present. In the rhombencephalon the rhombomeres are vanishing. The cerebellar plate has extended laterally, covering the lateral recesses of the fourth ventricle (Fig. 8B and C). At the dorsal junction between the rhombencephalon and the mesencephalon the decussation of the trochlear nerve is clearly visible.

3.5.4 Comparison with Other Mammals

The comparison of the characteristics analyzed in stage 17 embryos of different mammalian species is given in Table 6. Unfortunately, the Chinese hamster had to be

Fig. 9 A–F. Photomicrographs of stage 17 embryos. A, B and C. Horizontal sections; A and B embryo aged 35 ± 1 days; C embryo aged 34 ± 1 days p.c., D, E and F. Frontal sections of an embryo aged 36 ± 1 days p.c. Length of the *bars* noted in millimeters. Abbreviations: *CoD*, cochlear duct; *CoP*, cochlear pouch; *CP*, cerebellar plate; *DI*, diencephalon; *E*, epiphysis; *HY*, hypophyseal anlage; *LB*, lens body; *LNP*, lateral nasal process; *LVR*, lateral ventricular ridge; *MNP*, medial nasal process; *MP*, maxillary process; *MVR*, medial ventricular ridge; *NF*, nasal fin; *NL*, neural layer of retina; *PL*, pigment layer of retina; *SD*, semicircular duct; *TEL*, telencephalon; *VP*, vestibular pouch

Table 6. Characteristics of mammalian stage 17 embryos

	Man	Baboon	Rhesus monkey	Marmoset	Mouse
Age (days)	42–44	(33) 35±1 (39)	(32) 34–36 (38)	61–77	12
Length (mm)	11–14	10–13	9–12	±12.0	7–9
Cervical flexure < 90°	+	+	+	+	?
Thoracal somites vanishing	+	+	+	?	+
Definite nasomaxillary groove	+	+	+	+	+
Deep, medially located nasal pits	+	+	+	+	+
Auricular hillocks on mandibular bar	+	+	+	+	+
Retinal pigment visible	+	+	+	+	+
Subdivision within handplate	+	+	+	+	+
First sign of fingerrays	+	+	±	+	+
Subdivision of hindlimb bud	+	+	+	+	+
Footplate present	+	+	±	+	+
Lens cavity crescentic	+	+	+	+	+
Primitive lens fibers present	+	+	+	+	+
Lips optic fissure started fusion	+	+	+	?	?
Numerous retinal pigment granules	+	+	+	+	+
Primordial semicircular ducts	+	+	+	−	+
Primordial cochlear duct	+	+	+	−	+
Epithelia of medial and lateral nasal processes fused: nasal fin	+	+	+	+	+
Rathke's pouch closed	−	−	−		+
Neurohypophysis fingerlike projection of diencephalic floor	+	+	+	+	+
Expansion of telencephalic hemispheres	+	+	+	+	+
Epiphysis: small protrusion of diencephalic floor	+	?	+	?	+

omitted since the developmental stage 17 as presented by ten Donkelaar et al. (1979), which according to the authors resembles Carnegie stages 17 and 18, has most in common with stage 18 according to our criteria and will thus be included with that stage. As regards the mouse, Theiler (1972) equated his stage 20 to Streeter's (1948) horizon 17, thus the embryos described by Theiler are shown in Table 6.

On average the human stage 17 embryos appear to be about 8 days older than those of the baboon and the rhesus monkey, whereas the marmoset embryos are even twice as old as the latter. The mean length of the primate embryos analyzed shows no significant variance; no details are available on galago stage 17 embryos. In comparison with the mouse it can be noted that guinea pig stage 17 embryos are aged about 20 1/2 days, as can be deduced from the data given by Scott (1937). The lack of detailed information in the description on the development of the rabbit

(Edward 1968) precludes its inclusion beyond stage 16. In the rat, stage 17 embryos are probably about 14–14 1/4 days old and 9.4–10.3 mm long (Christie 1964).

The external characteristics of mammalian stage 17 embryos as listed in Table 6 prove the great conformity of these embryos. The lack of the first sign of fingerrays and the absence of a true, definite footplate were observed in the youngest member of the stage 17 rhesus monkey embryos. On the basis of its internal characteristics, however, the specimen had to be included in this group. Of the external characteristics mentioned the most notable are: auricular hillocks on the mandibular bar, the first sign of fingerrays, the subdivision of the hindlimb bud, and the presence of the footplate.

The external conformity of the stage 17 embryos is consolidated by their internal characteristics. The only remarkable difference is the fact that in the mouse Rathke's pouch is closed, as it was earlier in the Chinese hamster (stage 14), whereas in the primates it is still continuous with the stomodeum through a narrow opening. The most characteristic internal features of stage 17 embryos are: the crescentic lumen of the lens; the presence of primitive lens fibers and of numerous retinal pigment granules; and the neurohypophysis as a fingerlike projection of the diencephalic floor.

3.6 Stage 18

3.6.1 Postconceptional Age and Length

Altogether five rhesus monkey embryos representative of developmental stage 18 were examined. The age of these specimens varied between 35 ± 1 and 38 ± 1 days p.c. The crown-rump length of the embryos ranged from 11 to 15 mm.

3.6.2 External Features

In embryos at developmental stage 18 the cervical flexure has turned from an acute angle into an approximate right angle as the head has started to lift (Fig. 10A). The somites are still recognizable exteriorly in the lumbosacral region. Compared with the preceding stage the maxillary process is less protruding, while the heavily pigmented retina is more striking externally (e.g., Figs. 8A and 10A). A shallow frontonasal groove is present and the primitive nostrils are located medially, rather close together. The auditory groove is less narrow, forming the primitive external auditory meatus, which is surrounded by the six very prominent auricular hillocks (Fig. 10A).

The most conspicuous feature in the limbs is the presence of distinct fingerrays in the handplate. In the older members the rim of the handplate is crenated because of developing interdigital notches. The handplates are oriented about parallel to the midsagittal plane, thus with the future palmar surfaces of the hands facing medially (Fig. 10C). In the advanced members of the group, however, they have started to turn, thereby facing more mediocaudally. The hindlimb is definitely subdivided into a flattened, round footplate and a cylindrical leg segment (Fig. 10A). The orientation of the footplates is such that the future plantar surfaces are facing mediocranially (Fig. 10C). Toerays could not be recognized within the footplates. The tail of the embryos is still of considerable length and tapers into a fine curved tip (Fig. 10A–C).

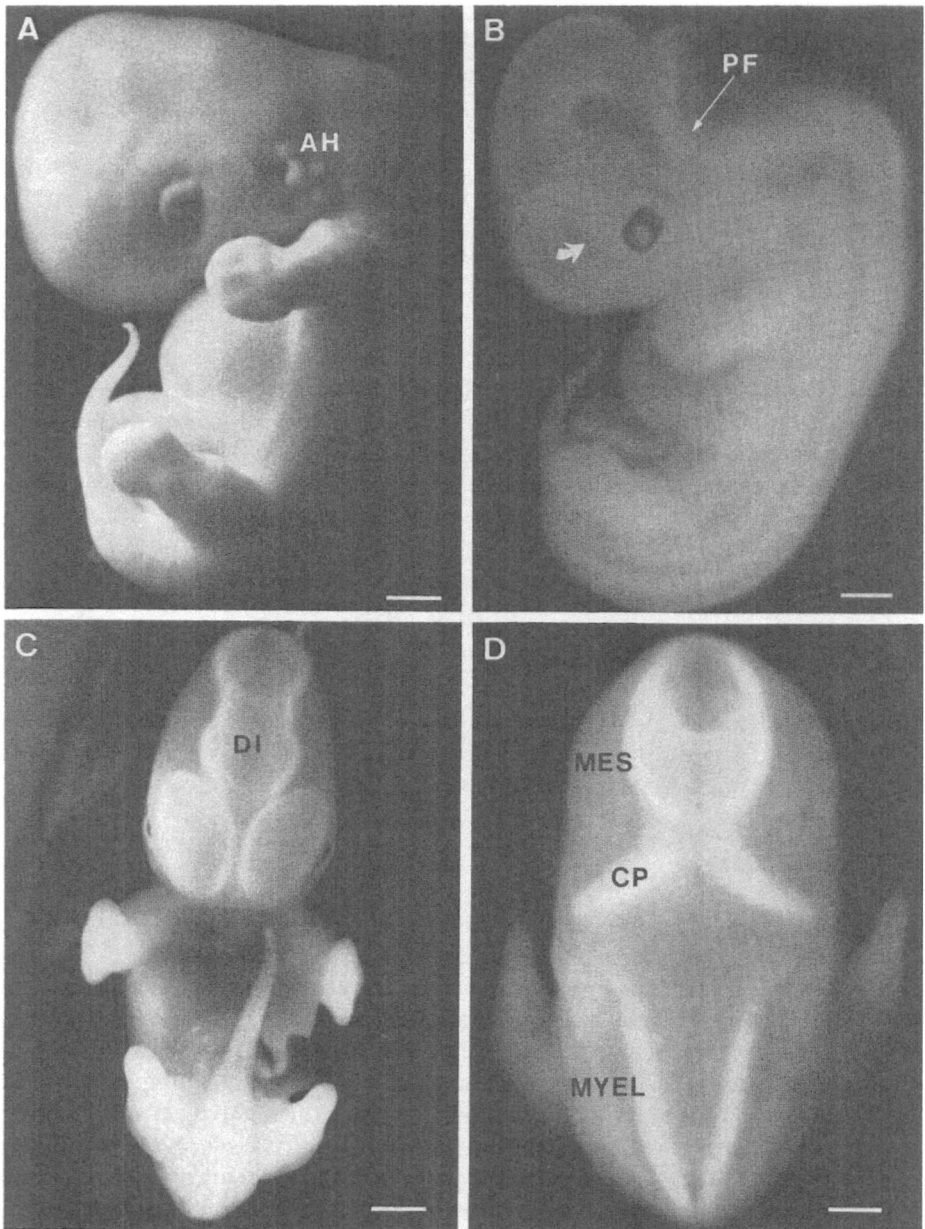

Fig. 10 A–D. Photomicrographs of stage 18 embryos. A and B. Lateral view of an embryo aged 36 ± 1 days p.c.; in B cleared in methylbenzoate, C and D. Ventral and superior views of an embryo aged 35 ± 1 days p.c. cleared in methylbenzoate. *Arrow* in indicates thickened basolateral part of the telencephalic hemisphere. Length of the *bars* 1 mm. Abbreviations: *AH*, auricular hillock; *CP*, cerebellar plate; *DI*, diencephalon; *MES*, mesencephalon: *MYEL*, myelencephalon; *PF*, pontine flexure

3.6.3 Internal Features

A characteristic feature of stage 18 embryos is the closure of the lens cavity (Fig. 11A), but in some places a small slit still remains. The inner thickened layer of the lens forming the lens body contains distinct primary lens fibers resulting from a progressive lengthening of its constituent cells. This latter process causes a gradual reduction of the lumen of the lens and eventually leads to its complete disappearance, thereby, bringing the outer epithelial layer of the lens into close contact with the lens body (Fig. 11A). The lens epithelium contains several rows of nuclei. The optic fissure is almost completely closed except for its most ventral part; the optic stalk contains a narrow lumen over its full length. The retinal pigment granules are arranged into a quadrangular cylinder (Fig. 10B) and are concentrating towards the inner surface of the pigment layer of the retina (Fig. 11A).

As regards the inner ear, a considerable elongation of the cochlear duct can be observed (Fig. 11B) in addition to the growth of the endolymphatic duct. Also, both the utriculus and sacculus become apparent, and all three semicircular ducts can be recognized.

The nasal pits have extended caudally but still end blindly, being separated from the oral cavity by a thin bucconasal membrane. The latter will be ruptured in the next stage. A small medial evagination of the nasal pit indicates the primordium of the vomeronasal organ of Jacobson. Within the septal area a condensation of mesenchymal cells preludes the primordium of the nasal septum cartilage. Caudally, the primitive tongue has started to project dorsally into the oral cavity, while the primordial palatine processes are recognizable within the primitive palate (Fig. 11D). Within the lower jaw condensations of mesenchymal cells are present, the primordia of Meckel's cartilages.

The neural hypophysis has extended more rostrally and the wall of its most rostral part has started to fold (Fig. 11C). The adenohypophysis has expanded dorsal to the neurohypophysis, resulting in the characteristic shape of the hypophyseal anlage in horizontal sections as illustrated in Fig. 11C. The opening of the adenohypophysis is closed, although in some specimens its lumen is still continuous with the oral cavity through a narrow canal (Fig. 11D).

In the CNS the angle of the flexura pontina becomes more and more acute as can be seen in the picture of an embryo cleared in methylbenzoate (Fig. 10B). In such embryos the thickened basolateral part of the telencephalic wall is easily recognizable. Compared with the preceding stage the frontal and caudal evagination of the telencephalic hemispheres has proceeded (e.g., Figs. 10B, C, and 8B, D). Also the cerebellar plates are much more impressive, which is clearly illustrated in Fig. 10D. From the microscopic sections it can easily be concluded that the thickened basolateral part of the telencephalic wall is composed of two separate elevations, the medial and lateral ventricular ridges (Fig. 11F). In the diencephalic roof a distinct commissura posterior is present (Fig. 11E) caudal to the epiphysis. The thin, undifferentiated roof of the fourth ventricle borders on the thick differentiating cerebellar plates (Fig. 11F).

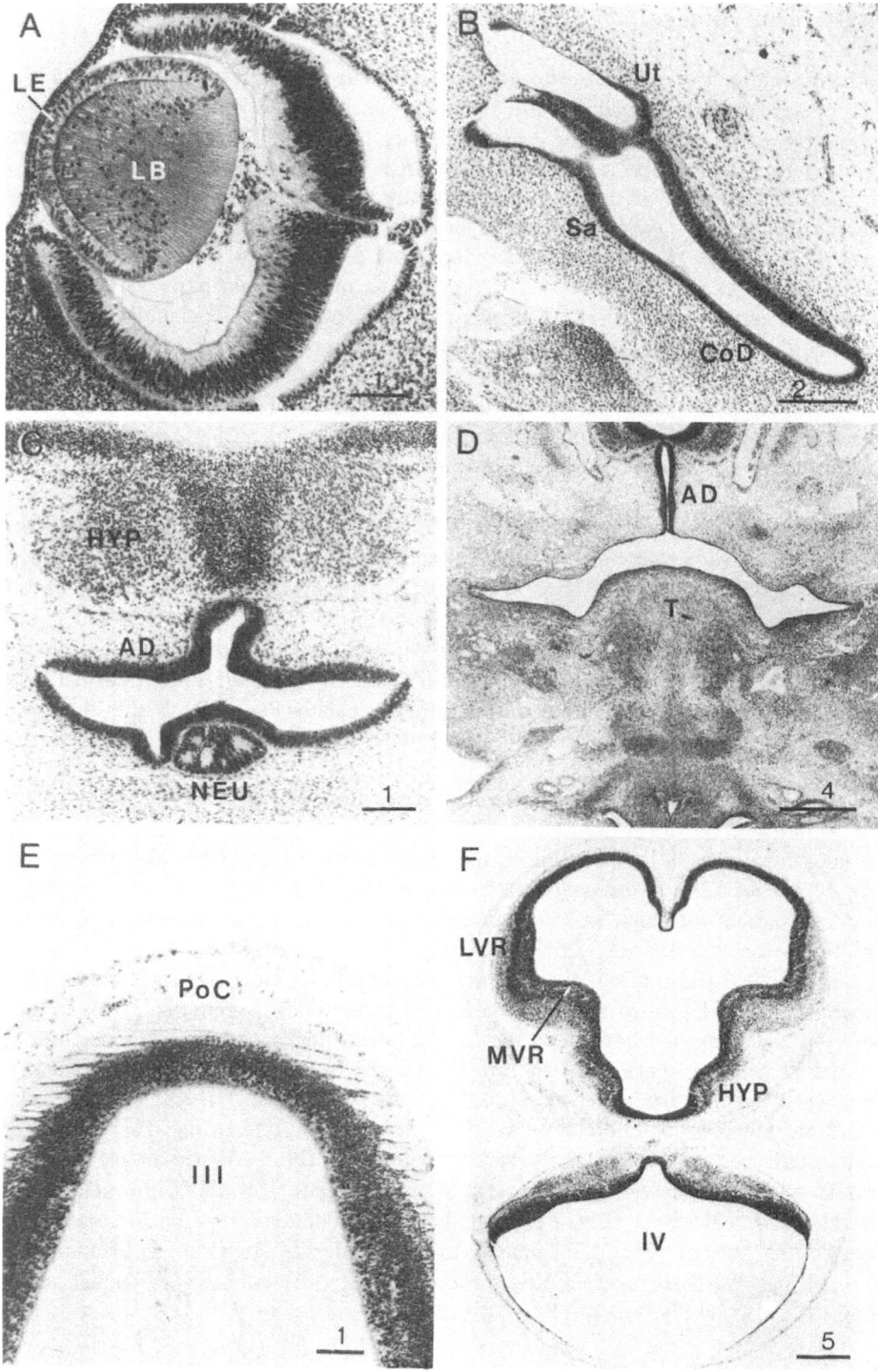

Fig. 11A–F

3.6.4 Comparison with Other Mammals

The comparison of mammalian stage 18 embryos is listed in Table 7. Unfortunately no data are available either on the marmoset (stage 18 was not described by Phillips 1976) or on the mouse (Theiler's stage 21 has more in common with developmental stage 19). As described in Sect. 3.5.4 before the Chinese hamster stage 17 as it was introduced by ten Donkelaar et al. (1979) is included here in stage 18 since the criteria used in the present investigation to define stage 18 correspond fairly well, as can be seen in Table 7. The human stage 18 embryos are about 9 days older than the respective specimens of both the baboon and the rhesus monkey, in which the embryos

Table 7. Characteristics of mammalian stage 18 embryos

	Man	Baboon	Rhesus monkey	Chinese hamster
Age (days)	44−48	(35) 37±1 (41)	(34) 35−38 (39)	14
Length (mm)	13−17	14−17	11−15	7.3−8.6
Cervical flexure circa 90°	+	+	+	+
Lumbosacral somites recognizable	+	+	+	?
Massive retinal pigmentation	+	+	+	+
Shallow frontonasal groove	+	+	+	+
Primitive nostrils	+	+	+	+
Wide external auditory meatus	+	+	+	+
Six prominent auricular hillocks	+	+	+	+
Distinct fingerrays	+	+	+	+
Rim of handplate crenated	±	+	±	±
Handplates facing mediocaudally	+	+	±	+
Definitive footplate	+	+	+	+
Footplate facing mediocranially	+	+	+	+
Lumen of lens cavity slitlike	+	+	+	+
Distinct primary lens fibers	+	+	+	+
Optic fissure almost closed	+	+	+	?
Optic stalk open	+	?	+	+
Bucconasal membrane ruptured	±	−	−	+
Primordial vomeronasal organ	+	−	+	?
Primordial nasal septum cartilage	+	+	+	+
Primitive palatine processes	+	+	+	+
Primordial Meckel's cartilages	+	+	+	?
Closure of Rathke's pouch	+	+	+	+
Started folding of neurohypophysis	+	+	+	?
Distinct epiphysis	+	?	+	+

Fig. 11 A−F. Photomicrographs of stage 18 embryos. A, B, C and F. Horizontal sections of an embryo aged 36 ± 1 days p.c., D. Frontal section of an embryo aged 35 ± 1 days p.c., E. Horizontal section of an embryo aged 35 ± 1 days p.c. Length of the *bars* noted in millimeters. Abbreviations: *AD*, adenohypophysis; *CoD*, cochlear duct; *HYP*, hypothalamus; *LB*, lens body; *LE*, lens epithelium; *LVR*, lateral ventricular ridge; *MVR*, medial ventricular ridge; *NEU*, neurohypophysis; *PoC*, posterior commissure; *Sa*, sacculus; *T*, tongue; *Ut*, utriculus; *III*, third ventricle; *IV*, fourth ventricle

are a little smaller than in the two other primates. The galago stage 18 embryo of un-known age as described by Butler (1972) measures 8.5 mm. In comparison with the Chinese hamster the guinea pig stage 18 embryos are probably about 21 1/2 days old (e.g., Scott 1937), and in the rat about 14 1/2 days and 11 mm (Christie 1964).

During the second half of the embryonic period in all mammalian species a gradual lifting of the head is seen toward the erect position which is present in the fetal period. In all species analyzed this process appears to start during developmental stage 18. Within the primates another process occurs in this phase of embryonic development namely the reduction of the tail. In human stage 17 embryos a clear short, tapering tail is present, whereas in stage 18 it is reduced to a blunt-ended process (cf. Streeter 1948). In baboon stage 18 embryos the tail reaches up to the level of the liver bulge, whereas in stage 17 embryos it ends near the forehead (cf. Hendrickx 1971). In the rhesus monkey embryos at stage 18 the tail still reaches up to the level of the forehead.

The most salient external features of mammalian stage 18 embryos are the presence of a wide external auditory meatus surrounded by six prominent auricular hillocks, the handplates showing distinct fingerrays and the presence of definite footplates.

Along with the great many internal similarities between primate and rodent stage 18 embryos some minor differences can be observed. Firstly, the bucconasal membrane is ruptured in the rodents, whereas in the primates it separates the oral and nasal cavities up to stage 19. Secondly, in primates the closure of Rathke's pouch occurs during stage 18, whereas in the mouse it is closed during stage 17, and in the Chinese hamster even earlier. The most distinctive internal features of the mammalian stage 18 embryos analyzed are the closure of the lens cavity, the presence of distinct primary lens fibers, and the onset of differentiation of various cartilages.

3.7 Stage 19

3.7.1 Postconceptional Age and Length

Eight rhesus monkey embryos examined were considered representative of developmental stage 19. Their postconceptional age ranged from 36 ± 1 to 42 ± 1 days. The crown-rump length of these embryos varied between 14 and 17 mm.

3.7.2 External Features

The overall external form of the embryos is dominated by the cervical flexure, which has increased to more than 90 °C (Fig. 12A). The distinct bulge at this flexure is caused by the underlying flexura cervicalis of the CNS (e.g., Fig. 12A and B). The spinal cord also protrudes exteriorly as a longitudinal column, while the somites are no longer visible.

Narrow ectodermal folds, the primordia of the eyelids, are present, while in the eyes a white opaque mass marks the condensation of scleral mesenchyme obscuring the retinal pigment exteriorly (Fig. 12A–C). The frontonasal fold has deepened, separating the nose from the forehead. The auricular hillocks, although all six are still recognizable, are less distinct than in the preceding stage. This results from their

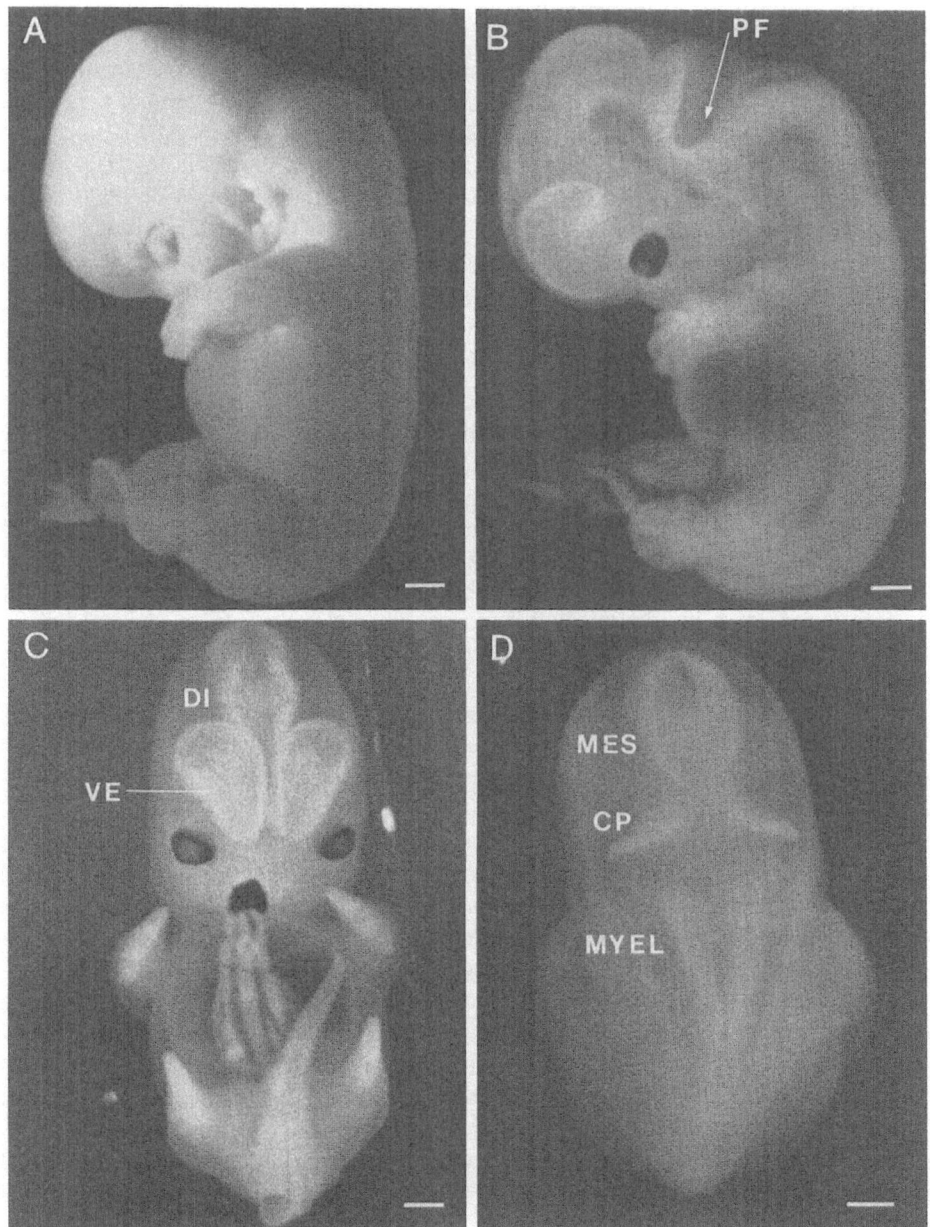

Fig. 12 A–D. Photomicrographs of a stage 19 embryo, aged 40 ± 1 days p.c. A and B. Lateral views, C. Frontal view, D. Superior view; in B, C and D cleared in methylbenzoate. Length of the *bars* 1 mm. Abbreviations: *CP*, cerebeller plate; *DI*, diencephalon; *MES*, mesencephalon; *MYEL*, myelencephalon; *PF*, pontine flexure; *VE*, ventricular eminence

incipient fusion to form the definite parts of the auricle; the external auditory meatus is still very wide (Fig. 12A).

The rim of the handplate is crenated due to the interdigital notches separating the individual fingerrays. The palmar surfaces of the hands have turned from a parasagittal position in the preceding stage into a caudomedial direction (e.g., Figs. 10C and 12C). The first sign of an elbow becomes recognizable (Fig. 12A and B). The shape of the footplates is pentagonal because of the toerays, which are definitely present at this stage. The footplates have rotated from a craniomedial into a parasagittal position, and their plantar surfaces now face medially (Fig. 12C). The tail of stage 19 embryos has become relatively shorter, now reaching up to the level of the umbilical cord.

3.7.3 Internal Features

The lumen of the lens cavity of stage 19 embryos is completely obliterated. The nuclei of the lens body are located in a lateral convex plane, as can be seen in the right lens in Fig. 13A, although the eyes of this specimen were damaged. The retinal pigment is arranged into a real cylinder as is visible in Fig. 12B and C. The optic stalk is closed except for its most proximal part near the diencephalon; the optic nerve fibers almost reach the latter part of the brain. The condensation of scleral mesenchyme under the surface extoderm is continuous with the mesenchymal cells between the lens epithelium and the ectoderm, the latter being the primordium of the cornea. The eyelids are blunt ectodermal folds.

In this and following developmental stages the inner ear can no longer serve as a determinant since its structure has become too complicated to allow an accurate interpretation from the microscopical sections without the use of reconstructive models.

The bucconasal membrane has invariably ruptured. As a consequence the nasal sac caudally communicates freely with the oral cavity through the primitive choanae (Fig. 13B). The vomeronasal organ of Jacobson is very distinct and the vomeronasal plexus is beginning to form. The mesenchymal condensation of the nasal septum cartilage has extended. The lateral palatine processes are very prominent structures lateral to the tongue while Meckel's cartilages are easily recognizable (Fig. 13B).

The adenohypophysis has lost its communication with the oral cavity. Within the adenohypophysis a pars distalis and a pars intermedia can be discerned (Fig. 13C). The neurohypophysis still exhibits a lumen continuous with the third ventricle, although in its rostral part the lumen is locally reduced because of the incipient folding of its wall.

In the CNS the outgrowth of the cerebral hemispheres in the frontal and caudal directions has proceeded (e.g., Figs. 10B, C and 12B, C). The most caudal part of the cerebral hemisphere partly covers the diencephalon (Fig. 12B). This picture of a methylbenzoate-cleared embryo also shows that the flexura pontina has progressively

---

Fig. 13 A–F. Photomicrographs of stage 19 embryos, aged 36 ± 1 days p.c. A, C and F. Frontal sections, B, D and E. Horizontal sections. Length of the *bars* noted in millimeters. Abbreviations: *ADd*, adenohypophysis pars distalis; *AD im*, adenohypophysis pars intermedia; *CP1*, choroid plexus; *DT*, dorsal thalamus; *E*, epiphysis; *HYP*, hypothalamus; *LPP*, lateral palatine process; *MC*, Meckel's cartilage; *NE*, nasal epithelium; *NEU*, neurohypophysis; *T*, tongue; *TEL*, telencephalon; *VE*, ventricular eminence; *VO*, vomeronasal organ; *VT*, ventral thalamus; *nl*, olfactory nerve; *III*, third ventricle; *IV*, fourth ventricle

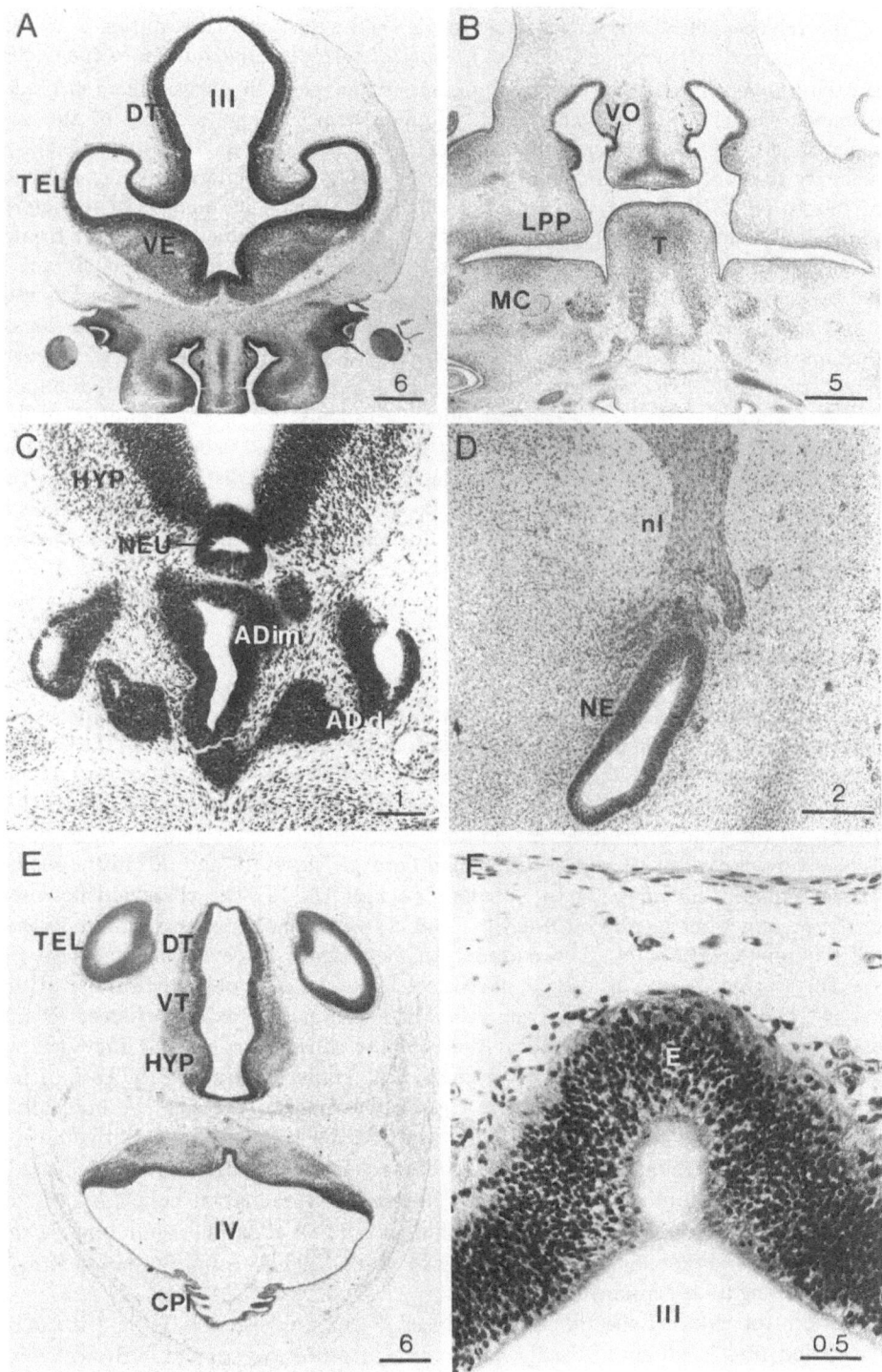

Fig. 13A–F

37

deepened compared with the preceding stage (Fig. 10B). The thickened basal parts of the telencephalic wall, recognizable in Fig. 12B and C, caudally appear to consist of one ventricular eminence (Fig. 13A), which is rostrally subdivided into the medial and lateral ventricular ridges already present in the preceding stage. The ventricular eminence apparently originates from a gradual process of merging of the two ventricular ridges which starts caudally (see also Lammers et al. 1980). The development of the ventricular eminence causes a narrowing of the interventricular foramen of Monro (Fig. 13A), and also the formation of the plexus choroideus of the lateral ventricle has started. The olfactory region is represented by a thickening of the frontal telencephalic wall and the first olfactory fibers coming from the nasal epithelium reach the brain (Fig. 13D). In the diencephalon the progress of the differentiation of its wall varies regionally: the hypothalamus and ventral thalamus are ahead of the dorsal thalamus (Fig. 13E). The epiphysis cerebri is a distinct evagination of the diencephalic roof (Fig. 13F); caudally the epiphysis is bordered by the broad fiber bundle of the commissura posterior. The fasciculus retroflexus can also be identified in this stage. In the mesencephalon the tectal part is very thin walled in contrast to the thickened tegmental part. In the cerebellar plates that are very distinct in the embryos cleared in methylbenzoate (Fig. 12B and D), the differentiation within their walls is obvious (Fig. 13E). The latter figure also documents the inception of the formation of the plexus choroideus of the fourth ventricle.

3.7.4 Comparison with Other Mammals

The characteristics of the embryos of various mammalian species representative of developmental stage 19 are listed in Table 8. It has to be mentioned that in the Chinese hamster and in the mouse those stages are included which were numbered as stage 18 and stage 21 respectively by the authors in question (ten Donkelaar et al. 1979; Theiler 1972). In the Chinese hamster the authors regarded their stage 18 as comparable to Theiler's mouse stage 21 and as resembling Carnegie stages 19 and 20. In the mouse Theiler equated his stage 21 to Streeter's horizon 18–19. The characteristics used in the present investigation, as listed in Table 8, provide evidence that in both species the embryos described by these authors can be assigned to developmental stage 19.

On average, human stage 19 embryos are about 10 days older than those of the baboon and the rhesus monkey, and also a little longer. The marmoset stage 19 embryos are still about twice as old as those of the other primates, but they are obviously smaller. The galago stage 19 embryo described by Butler (1972) is of unknown age and measures 11 mm. As regards the rodents, it should be noted that developmental stage 19 is attained 2 days later in the Chinese hamster than in the mouse, as in the previous stages, but the embryos show only a slight variation in length. From the description of rat embryos by Christie (1964) it may be concluded that in this species stage 19 embryos are about 15 days old and about 11.5 mm long. In the study of the embryology of the guinea pig by Scott (1937) no embryo was described corresponding to developmental stage 19.

From the external characteristics of stage 19 embryos listed in Table 8 it can be concluded that in all species analyzed the outer features correspond fairly well with only minor variations. Compared with the preceding stage the process of lifting of the head has advanced: the cervical flexure now measures more than 90°. Within the

Table 8. Characteristics of mammalian stage 19 embryos

	Man	Baboon	Rhesus monkey	Marmoset	Chinese hamster	Mouse
Age (days)	48–51	(38) 39±1 (40)	(35) 36–42 (43)	±75	$15-15^{1/4}$	13
Length (mm)	16–18	16–17	14–17	12.1–12.4	8.5–10	9–11
Cervical flexure > 90°	+	+	+	+	+	+
Somites absent	+	+	+	+	+	+
Cervical bulge present	+	+	+	?	+	+
Eyelids: small ectodermal folds	+	+	+	+	+	+
Deep frontonasal groove	+	+	+	+	+	+
Auricular hillocks start fusion	+	+	+	+	+	+
Handplates facing caudomedially	+	+	+	+	+	+
Rim of handplate crenated	+	+	+	+	+	+
First sign of elbow	+	+	+	+	?	+
Distinct toerays	+	+	+	+	+	+
Footplates facing medially	±	+	+	±	±	+
Rim of footplate crenated	±	±	±		+	+
Lens cavity obliterated	+	+	+	+	+	+
Optic stalk closed	+	+	+	+	+	+
Primordial cornea present	+	+	+	+	+	+
Eyelids are blunt ectodermal folds	+	+	+	+	+	+
Bucconasal membrane ruptured	+	+	+	+	+	+
Distinct vomeronasal organ	+	±	+	+	?	+
Nasal septum cartilage distinct	+	+	+	?	?	+
Meckel's cartilages distinct	+	+	+	?	+	+
Palatine processes bilateral to the tongue	+	+	+	+	+	+
Subdivision of adenohypophysis	+	+	+	+	?	+
Lumen neurohypophysis continuous with 3rd ventricle	+	+	+	+	+	+
Formation of plexus choroideus of lateral and fourth ventricles	+	?	+	?	?	+

primates the relative shortening of the tail is now also apparent in the rhesus monkey: its tip now reaches the level of the umbilical cord as it does in the baboon (cf. Hendrickx 1971), whereas in the human embryos at this stage only a stump is present (cf. Streeter 1951). In all species analyzed during developmental stage 19 the auricular hillocks start to fuse, which ultimately leads to the formation of the auricula. For detailed information on the development of the auricle the reader is referred to Streeter (1922). Another process that is evident during this phase of development is the rotation of the extremities: the palmar surfaces of the hands turn from a medial into a caudal position and the plantar surfaces of the feet turn from cranially toward a medial position.

The main discriminating external features compared with the preceding stage are: the cervical flexure is more than 90°; the auricular hillocks have started to fuse; the palmar surfaces of the hands face more caudally than medially; the rim of the handplate is crenated; distinct toerays are present; and the plantar surfaces of the feet face more medially than caudally.

Regarding the internal characteristics of stage 19 embryos of the various mammalian species listed in Table 8 hardly any variations can be noticed, although a few of the items were not analyzed by Hendrickx, Phillips and ten Donkelaar. The most salient features of stage 19 embryos in comparison with those of stage 18 are: the lens cavity is obliterated; the optic stalk is closed; the bucconasal membrane is ruptured; distinct Meckel's cartilages are present, and definitive processus palatini project lateral to the tongue.

3.8 Stage 20

3.8.1 Postconceptional Age and Length

The postconceptional age of the five embryos at developmental stage 20 varied between 38 ± 1 and 42 ± 1 days. One embryo aged 36 ± 1 days and one aged 44 ± 1 days p.c. had to be included in stage 20 on account of their internal and external characteristics. The crown-rump length of the embryos ranged from 16 to 20 mm.

3.8.2 External Features

The angle of the cervical flexure of stage 20 embryos has increased up to circa 120° by stage 20 due to the progressive lifting of the head (Fig. 14A). The flexura cervicalis of the CNS still causes an external bulge, while the spinal cord also protrudes exteriorly as a longitudinal column although both phenomena are less impressive than in the preceding stage 19 (e.g., Figs. 14A, B and 12A, B).

The outer form of the head is almost quadrangular with a slight protrusion of the forehead, that results in a deepening of the frontonasal fold. The ectodermal folds of the primitive eyelids have started to grow and now cover the upper and lower parts of the eye (Fig. 14A). The first sign of primordial hair follicles can be recognized at the site of the future eyebrows. Definite auricles are formed that cover about half of the external auditory meatus.

The arms and legs are almost parallel, approximately at right angles to the dorsum of the embryo (Fig. 14A). Slight flexures are present indicating both the elbows and the

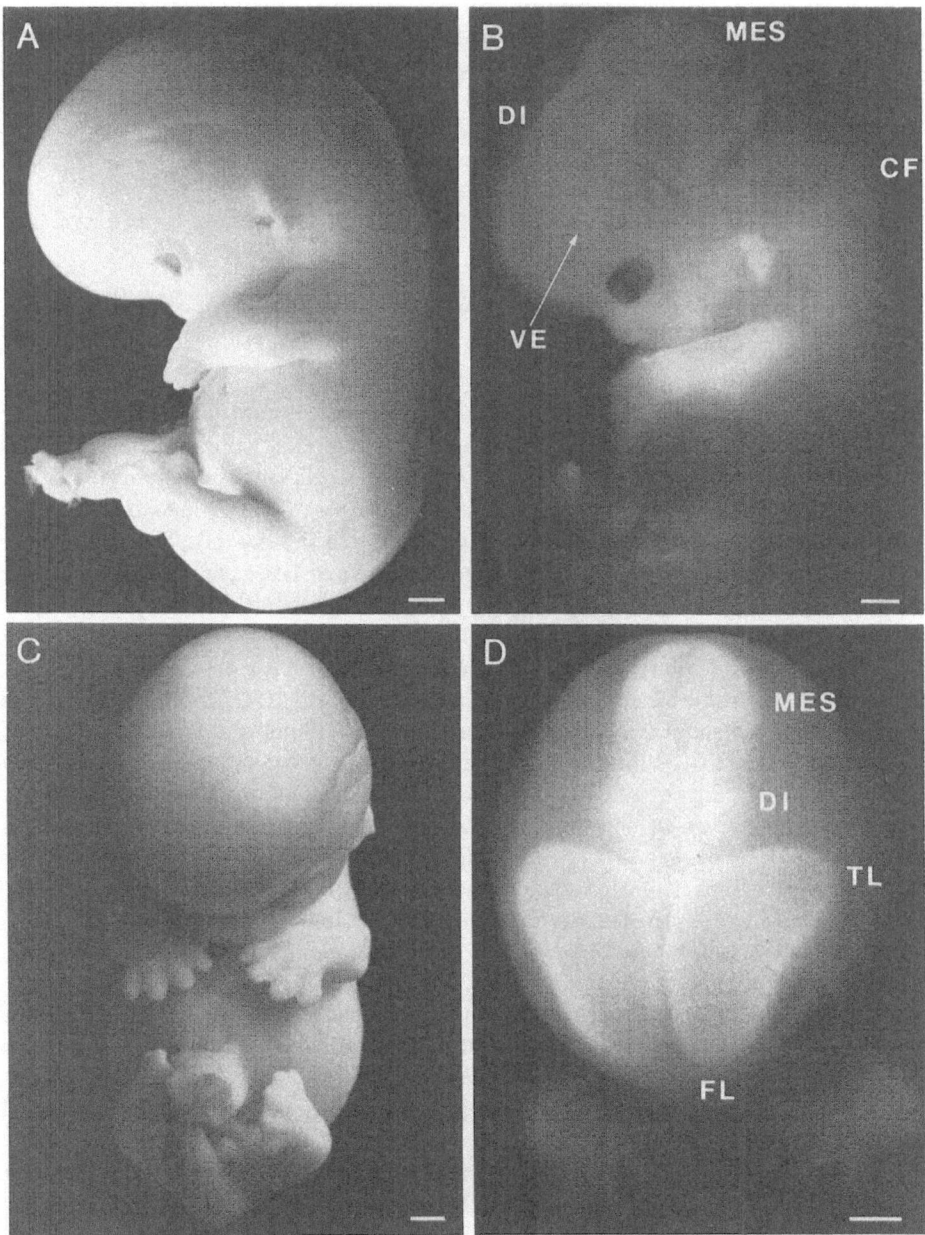

Fig. 14 A–D. Photomicrographs of stage 20 embryos. A and C. Lateral and ventral views of an embryo aged 40 ± 1 days p.c., B and D. Lateral and superior views of an embryo aged 38 ± 1 days p.c. cleared in methylbenzoate. Length of the *bars* 1 mm. Abbreviations: *CF*, cervical flexure; *DI*, diencephalon; *FL*, frontal lobe; *MES*, mesencephalon; *TL*, temporal lobe; *VE*, ventricular eminence

41

knees. The hands are located ventromedially, and their rotation has advanced; the palmar surfaces now face caudally. A characteristic feature of developmental stage 20 is the onset of separation of the fingers: the individual fingertips are easily recognizable (Fig. 14C). The growth of the legs is attended by a more ventromedial position of the footplates, which are oriented parallel to the midsagittal plane, their plantar surfaces facing medially (Fig. 14C). The toerays are very prominent and the rim of the foot-plate is crenated due to the interdigital notches (Fig. 14A and C). The length of the tail is about the same as in the preceding stage; it ends in a minute tip at a level near the umbilical cord.

3.8.3 Internal Features

The eyelid folds have become less blunt, and now cover about one-fifth of the eye surface (Fig. 15A, B and D). The lens is a solid sphere, consisting of primordial lens fibers with scattered nuclei, its outer part covered by a thin lens epithelium (Fig. 15B). Between the primordial cornea and the lens the future pupillary membrane can be identified. The lentoretinal space is developing (Fig. 15B, and D). The neural layer of the retina can be subdivided into an outer, nucleated layer and an inner, cell-free marginal layer (Fig. 15B). The optic nerve fibers have reached the diencephalon where the optic chiasm becomes recognizable.

Within the oral cavity the lateral palatine processes have extended. Posteriorly, they project vertically on either side of the tongue (Fig. 15D). Anteriorly, the lateral palatine processes show a medial extension that projects under the lateral surface of the tongue (Fig. 15A). Within the nasal septum cartilage the onset of differentiation of chondroblasts can be recognized, while the cartilages of the alae majores of the nose have also started to develop (Fig. 15A). The vomeronasal organ of Jacobson is not discernable as it was in the preceding stage. Olfactory fibers coming from the nasal epithelium form the primordial olfactory nerve (Fig. 15A and C).

The neurohypophysis has extended frontally; its lumen (not visible in Fig. 16A) is reduced but is still continuous with the third ventricle. The lumen of the adenohypophysis remained rather wide. The constituent cells of the pars distalis of the adenohypophysis have increased markedly in number, especially in its frontal part. Some secondary vesicles are formed, their lumen being continuous with the original lumen (Fig. 16A). The presence of some mesodermal capillaries indicates the onset of vascularization of the adenohypophysis. The pars intermedia has retained its thin-walled character.

Within the CNS of stage 20 embryos a progressive outgrowth of the cerebral hemispheres in the frontal, caudal, and lateral directions can be seen in the specimens cleared in methylbenzoate (e.g., Fig. 14B, D and 12B, C). The first process is probably implicated in the protrusion of the forehead (Fig. 14A–D). The latter outgrowth results in the formation of the temporal lobes, as is visualized in Fig. 14B and D. The cerebellar plates almost completely cover the lateral recesses of the fourth ventricle (Fig. 14B).

From the microscopic sections of the brain it can be seen that the choroid plexuses project far into the lumen of the lateral ventricles (Fig. 16C and D). The ventricular eminence (visible in Fig. 14B), caudally consists of one single massive elevation (Fig. 16C). Frontally, however the medial and lateral ventricular ridges still exist,

42

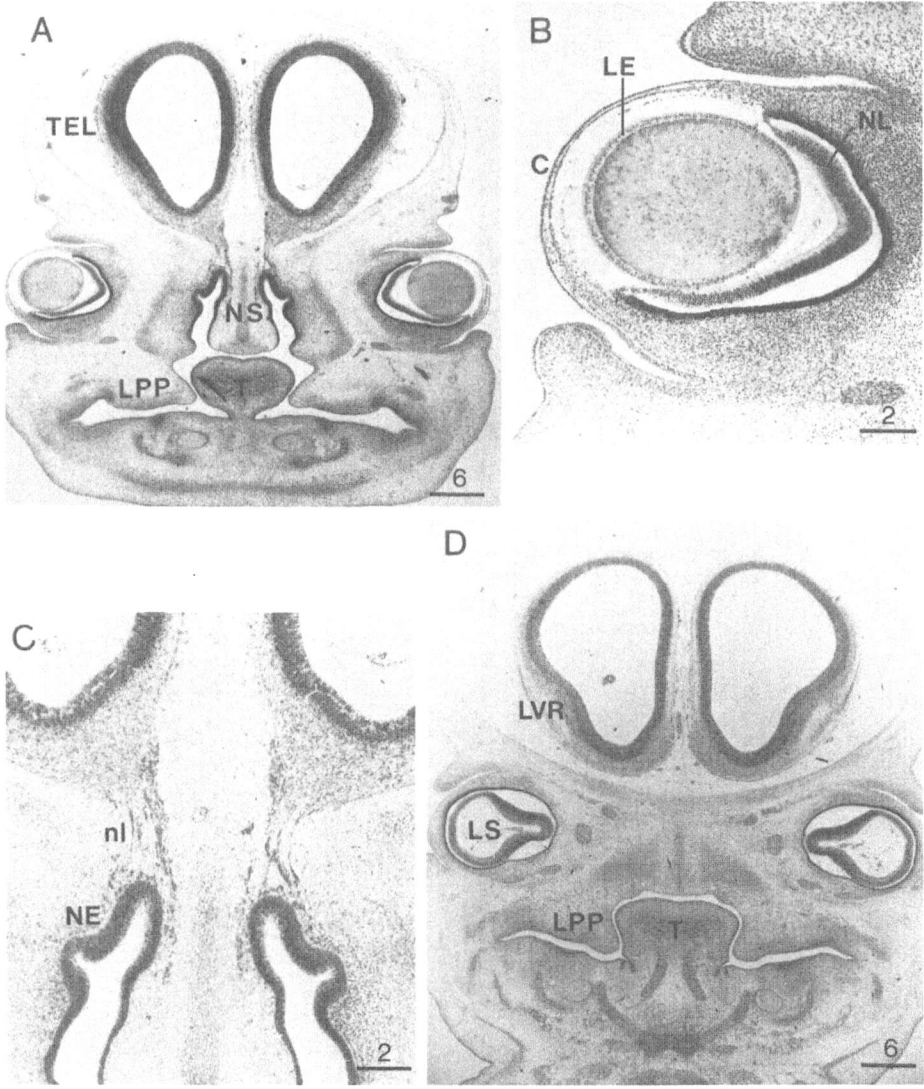

Fig. 15 A-D. Photomicrographs of a stage 20 embryo aged 40 ± 1 days p.c., horizontal sections. Length of the *bars* noted in millimeters. Abbreviations: *C*, cornea; *LE*, lens epithelium; *LPP*, lateral palatine process; *LS*, lentoretinal space; *LVR*, lateral ventricular ridge; *NE*, nasal epithelium; *NL*, neural layer of retina; *NS*, nasal septum; *T*, tongue; *TEL*, telencephalon; *nI*, olfactory nerve

separated by a rather deep sulcus (Fig. 16D). In the outer part of the telencephalic wall a superficial cortical layer (primary cortex) can be recognized, consisting of darkly stained cells (Fig. 16D). The internal capsule (not illustrated in Fig. 16) has become a distinct fiber mass.

The lumen of the third ventricle has narrowed, especially in its ventral part (Fig. 16C). Frontally, the diencephalic roof has started the differentiation leading

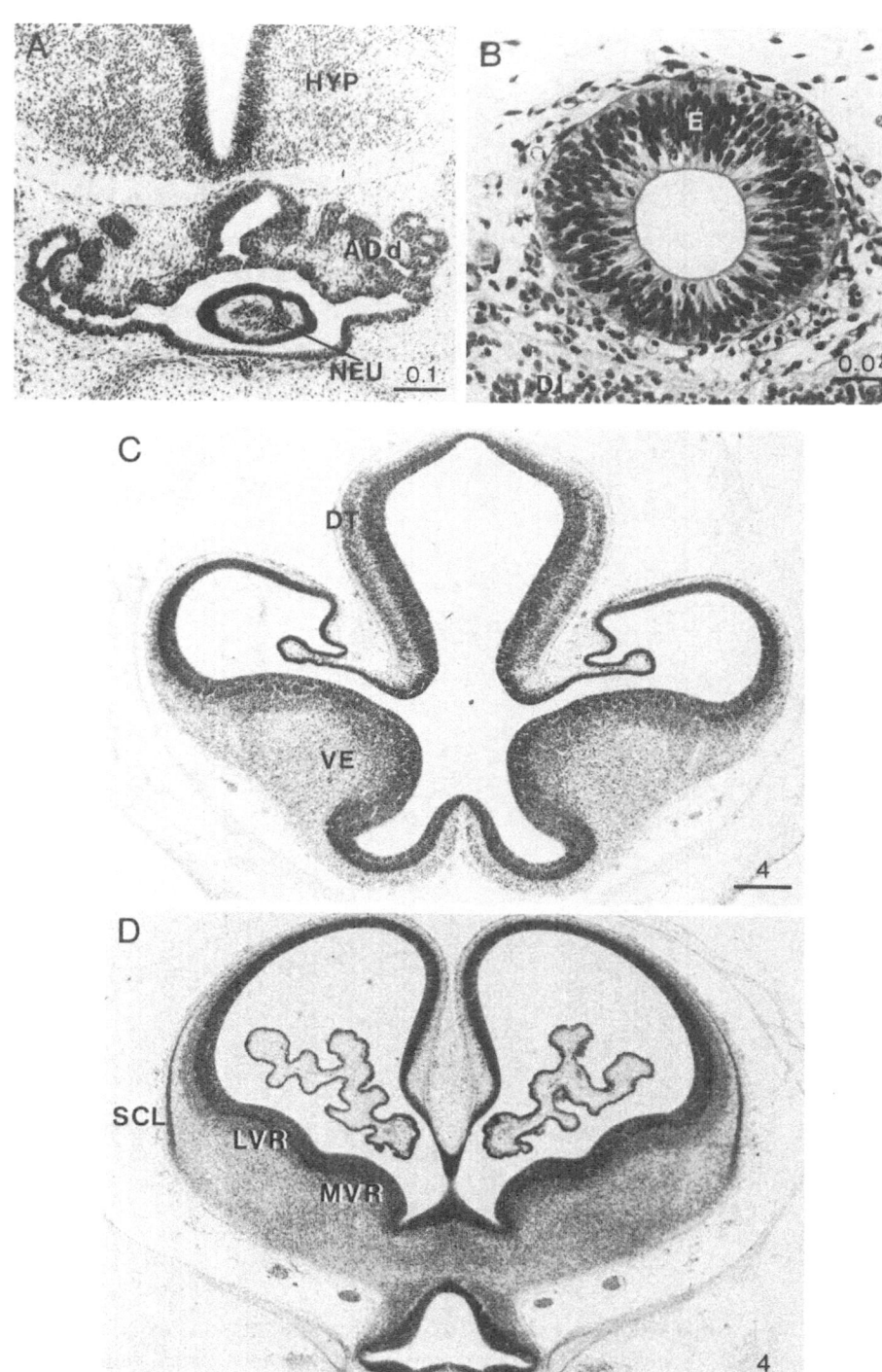

Fig. 16A–D

to the formation of a plexus choroideus, caudally it shows the rostrally directed fingerlike projection of the epiphysis (Fig. 16B). The lumen of the epiphysis is continuous with the third ventricle, at its ventricular surface some mitotic figures are present (Fig. 16B). Caudally, the epiphysis borders upon the broad posterior commissure. The infundibular recess is continuous with the narrowed lumen of the neurohypophysis. The optic chiasm can readily be identified. In the mesencephalon the tegmental part of the wall gradually thickens while the tectal part retains its thin-walled character. The cerebellar plates have extended and the choroid plexus projects far into the lumen of the fourth ventricle.

3.8.4 Comparison with Other Mammals

As described in the introductory remarks of Section 3 a detailed comparison of primate and rodent stages 20 proved to be impracticable. Thus the data presented in Table 9 are restricted to primates for no detailed information is available upon rodent embryos at a level comparable to stage 20. Information on the marmoset is also lacking since Phillips (1976) had no embryos representative of developmental stage 20 at his disposal.

From Table 9 it can be deduced that human embryos representative of stage 20 are on average 11 days older than those of the baboon and 12 days older than those of the rhesus monkey. They are also about 2 mm longer than those in the latter two species. The rat embryos described by Christie (1964), which on the basis of the onset of the separation of the fingers and of the medial outgrowth of the lateral palatine processes are comparable to stage 20 embryos of primate species, are aged 15 1/2–16 days, and measure 12.1–12.7 mm.

There are two externally visible processes that apparently start during developmental stage 20 and proceed during later stages: first the separation of the fingers and second the outgrowth of the eyelids. According to Christie (1964), however, in the rat the latter process only starts during developmental stage 21. On the other hand, he mentions the appearance of trunk hair papillae as a distinguishing feature of stage 20.

The most characteristic external features of primate stage 20 embryos are: the eyelids start growing; a definitive auricula is present; the distal fingertips are separated; and the rim of the footplate is crenated.

Apart from the absence of a vomeronasal organ in the rhesus monkey, as can be seen in Table 9, the internal characteristics of stage 20 primate embryos are identical. The most discriminative internal features in comparison with the preceding stage 19 are: the lens is a solid sphere, in which primordial lens fibers are present; the lateral palatine processes, which posteriorly are vertically oriented, show a medial extension anteriorly; the pars distalis of the adenohypophysis, in which an incipient vascularization is present, has started to differentiate.

Fig. 16 A–D. Photomicrographs of stage 20 embryos. A and D. Horizontal sections of an embryo aged 40 ± 1 days p.c., B and C. Frontal sections of an embryo aged 38 ± 1 days p.c. Length of the *bars* noted in millimeters. Abbreviations: *ADd*, adenohypophysis pars distalis; *DI*, diencephalon; *DT*, dorsal thalamus; *E*, epiphysis; *HYP*, hypothalamus; *LVR*, lateral ventricular ridge; *MVR*, medial ventricular ridge; *NEU*, neurohypophysis; *SCL*, superficial cortical layer; *VE*, ventricular eminence

Table 9. Characteristics of primate stage 20 embryos

	Man	Baboon	Rhesus monkey
Age (days)	51–53	(39) 41±1 (42)	(35) 38–42 (45)
Length (mm)	18–22	17–18	16–20
Cervical flexure circa 120°	+	+	+
Cervical bulge present	+	+	+
Eyelids start growing	+	+	+
Primordial hair follicles at eyebrows	?	?	+
Definite auricula present	+	+	+
Arms and legs about parallel	+	+	+
Slight flexures at elbows and knees	+	+	+
Palmar surfaces of hands face caudally	+	+	+
Distal fingertips separated	+	+	+
Footplates about parallel to median plane	+	+	+
Plantar surfaces of feet face medially	+	+	+
Prominent toerays	+	+	+
Rim of footplate crenated	+	+	+
Lens: solid sphere	+	+	+
Primordial lens fibers present	+	+	+
Subdivision of retina into two layers	+	+	+
Optic nerve fibers reach diencephalon	±	±	+
Lat. palat. processes: posteriorly vertical, anteriorly medial extension	+	+	+
First chondroblasts in nasal septum	?	+	+
Vomeronasal organ present	+	+	–
Pars distalis of adenohypophysis: first secondary vesicles	+	+	+
Adenohypophysis: beginning vascularization	+	+	+
Neurohypophysis: lumen continuous with 3rd ventricle	+	+	+

3.9 Stage 21

3.9.1 Postconceptional Age and Length

In total four rhesus monkey embryos belonging to developmental stage 21 were available. The postconceptional age of all four specimens varied between 40 ± 1 and 44 ± 1 days. Their crown-rump length ranged from 18 to 22 mm.

3.9.2 External Features

In the stage 21 embryos the neck region becomes apparent, as can be seen in lateral pictures (Fig. 17A). In frontal views (Fig. 17C) the face is readily visible, because of the continued elevation of the head. The outer form of the head has changed markedly compared with stage 20 embryos (e.g., Figs. 17A and 14A). This is probably due to the disappearance of the cervical bulge and the further frontal protrusion of the forehead. Also the spinal cord is no longer recognizable on the outside.

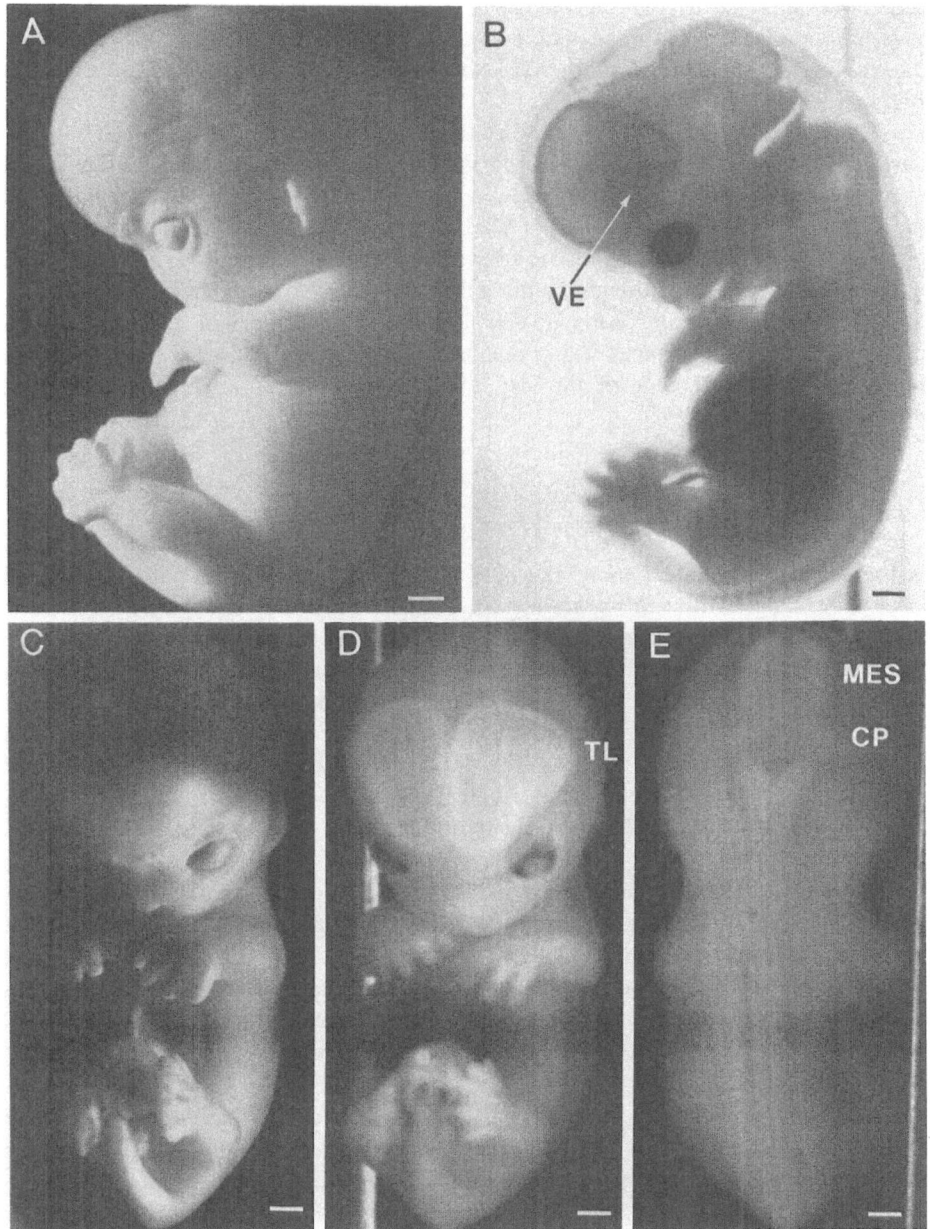

Fig. 17 A–E. Photomicrographs of a stage 21 embryo, aged 42 ± 1 days p.c. A and C. Lateral and ventral views of the embryo. B, D and E. Lateral, ventral, and dorsal views of the embryo cleared in methylbenzoate. Length of the *bars* 1 mm. Abbreviations: *CP*, cerebellar plate; *MES*, mesencephalon; *TL*, temporal lobe; *VE*, ventricular eminence

The growth of the eyelids has proceeded; they now cover about one-fourth of the external surface of the eyes (Fig. 17A). At the eyebrows distinct hair follicles are present (Fig. 17A and C). The auricula projects laterally, and the primordial helix and anthelix become recognizable. Compared with the preceding stage the jaws have elongated, the upper jaw extending further than the lower jaw; however, the volume of the neurocranium far exceeds that of the viscerocranium (e.g., Fig 17A and B).

The arms and legs have lengthened and no longer extend straight forward, but show obvious flexures at the elbows and the knees (Fig. 17A). Also the first sign of the wrists becomes visible; the hands show a palmar flexed position. The separation of the fingers has proceeded; the distal phalanges are separated. The individual fingers of the left and right hands almost touch each other in front of the snout (Fig. 17C). In the older members of the group the ankles become also recognizable. The position of the footplates has not changed; they are still oriented parallel to the midsagittal plane, their plantar surfaces facing medially. The footplates draw nearer, probably because of the lengthening of the legs. The toes have started to separate the tips of the individual toes are free.

3.9.3 Internal Features

In the microscopical sections of the eyes of stage 21 embryos the wide lentoretinal space is striking. Within this space a primordial vitreous body has started to develop (Figs. 18A and B). Within the lens definitive lens fibers become apparent. The anterior chamber is present between the cornea and the pupillary membrane. Within the cornea three component parts can be discerned: the epithelium, the substantia propria, and the mesothelium. At the retina-optic nerve junction fibers can be seen leaving the retina toward the optic nerve (Fig. 18A). The optic nerve is recognizable in its full extent up to the diencephalon, where many fibers cross in the optic chiasm (e.g., Fig. 19D). Underneath the pigmental layer of the retina a scleral condensation is present (Fig. 18A and B).

The lumen of the oronasal cavity has considerably decreased because of the progressive elevation of the tongue (Fig. 18C, D, E, and F). Posteriorly, the lateral palatine processes point vertically bilateral to the tongue (Fig. 18D). The most frontal parts of the lateral palatine processes are apposed to the nasal septum (Fig. 18F) and are even fused with the median nasal process or primary palate (Fig. 18C). The differentiation of the cartilages of both the nasal septum and of the alae majores has proceeded (Fig. 18D–F).

Within the hypophysis the progressive differentiation is clearly illustrated when Fig. 19A is compared with Fig. 16A. The pars intermedia of the adenohypophysis shows frontally a pair of small lobes bilateral to the neurohypophysis (Fig. 19A). More caudally these projections appear to be continuous with the pars infundibularis, which also extends dorsal to the neurohypophysis (Fig. 19D). The pars distalis of the adenohypophysis shows many strands of cells or trabeculae projecting from its original wall (Fig. 19A). Numerous trabeculae form secondary vesicles having a lumen that is continuous with the original lumen. The vascular component between the trabeculae and vesicles has markedly increased compared with the preceding stage. The neurohypophysis has hardly changed, it is still a frontal extension of the diencephalic floor with a small lumen (Fig. 19A and D) continuous with the third ventricle.

Except for its growth and the deepening of the pontine flexure the total form of the CNS has hardly changed compared with the preceding stage, as can be seen in the

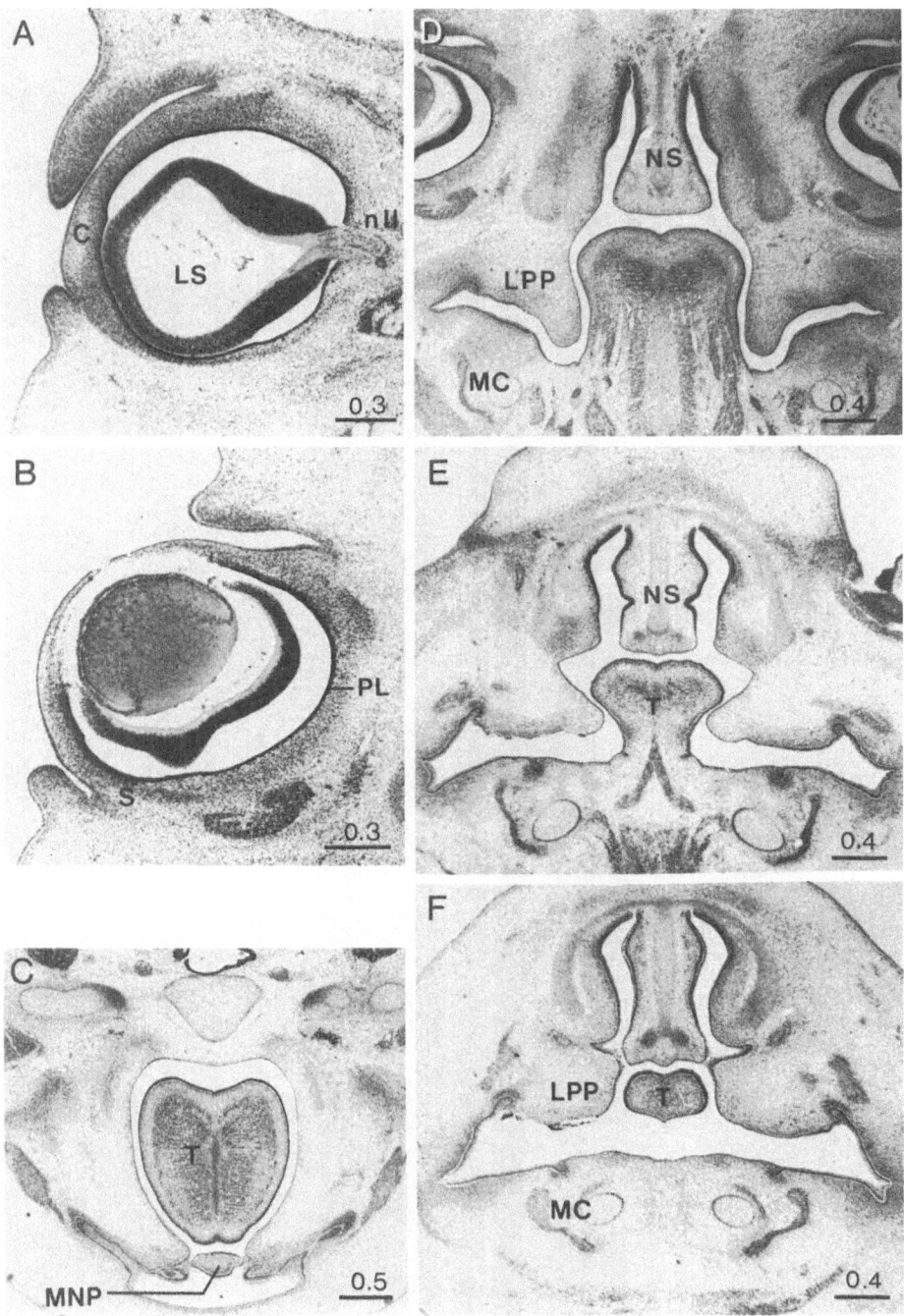

Fig. 18 A–F. Photomicrographs of stage 21 embryos. A, B, D, E, and F. Horizontal sections of an embryo aged 42 ± 1 days p.c.; C. Frontal section of an embryo aged 44 ± 1 days p.c. Length of the *bars* noted in millimeters. Abbreviations: *C*, cornea; *LPP*, lateral palatine process; *LS*, lentoretinal space; *MC*, Meckel's cartilage; *MNP*, medial nasal process; *NS*, nasal septum; *PL*, pigment layer of retina; *S*, sclera; *T*, tongue; *n II*, optic nerve

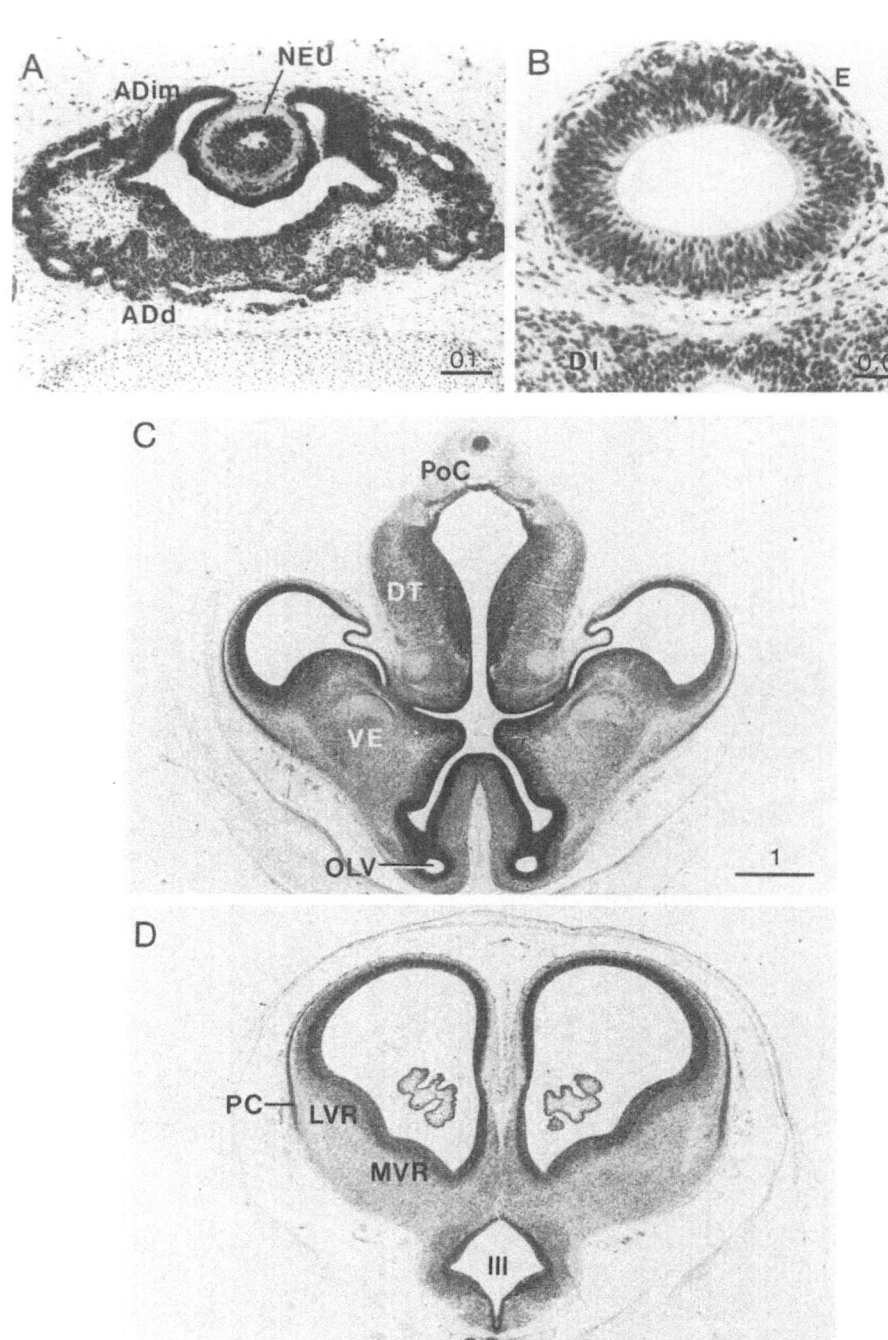

Fig. 19A–D

pictures of the embryos cleared in methylbenzoate (e.g., Figs. 17B and 14B). The outgrowth of the frontal and temporal lobes of the hemispheres has proceeded (Fig. 17B and D). Thereby both the frontal and dorsal parts of the diencephalon are covered by the telencephalon. In the embryos cleared with methylbenzoate the ventricular eminences are recognizable from the outside (Fig. 17B). From the sections of the CNS it can be concluded that the merging of the medial and lateral ventricular ridges, which started caudally, has proceeded frontally. Nevertheless, the frontal parts of the ventricular ridges remain separated by a rather deep groove (Fig. 19C and D). The superficial cortical layer has thickened, now comprising up to seven layers of cells. The olfactory ventricle is rather wide (Fig. 19C) and is still continuous with the lateral ventricle.

The third ventricle has progressively narrowed, especially in its thalamic part (Fig. 19C) and still shows a continuation into the neurohypophysis. The epiphysis remains a fingerlike projection of the diencephalic roof, having a rather wide circular lumen (Fig. 19B). Numerous mitoses can be observed in the pineal gland: they occur exclusively at its ventricular surface. Also the first sign of differentiation of cells forming mammilliform projections can be recognized. The epiphysis is bordered caudally by the commissura posterior (Fig. 19C). The optic chiasm has now formed a definite fiber mass (Fig. 19D).

A dorsal view of the caudal part of the CNS in stage 21 embryos is presented in Fig. 17E. The mesencephalon is situated above the cerebellar plates, which in turn hang over the myelencephalon (see also Fig. 17B). The thin roof permits a view of the fourth ventricle. The parts of the CNS under consideration do not exhibit any striking alterations except for a thickening of their basal areas.

3.9.4 Comparison with Other Mammals

The data available on the characteristics of mammalian stage 21 embryos are summarized in Table 10. The information on stage 21 marmoset embryos given by Phillips (1976) was not sufficient to include that species in Table 10. The age of these marmoset embryos ranged from 73 to 87 days and their length varied between 14.4 and 15.8 mm. No data are available on Chinese hamster embryos comparable to our stage 21. In the mouse, Theiler (1972) equated his stage 22 with Streeter's (1951) horizons 20–23. From the list of criteria presented in Table 10 it can be concluded that Theiler's stage 22 is highly comparable to our stage 21.

The trend of the difference in age of the human embryos compared with the other primates continues: the human embryos are about 10–12 days older than those of the baboon and the rhesus monkey. The same is true for their respective lengths; the human embryos are about 3 mm longer than those in the other two primate species. Compared with the mouse stage 21 embryos it should be mentioned

Fig. 19A–D. Photomicrographs of stage 21 embryos. A, B and C. Frontal sections of an embryo aged 44 ± 1 days p.c., D. Horizontal section of an embryo aged 42 ± 1 days p.c. Length of the *bars* noted in millimeters. Abbreviations: *ADd*, adenohypophysis pars distalis; *AD im*, adenohypophysis pars intermedia; *DI*, diencephalon; *DT*, dorsal thalamus; *E*, epiphysis; *LVR*, lateral ventricular ridge; *MVR*, medial ventricular ridge; *NEU*, neurohypophysis; *OLV*, olfactory ventricle; *PC*, primary cortex; *PoC*, posterior commissure; *VE*, ventricular eminence; *III*, third ventricle

Table 10. Characteristics of mammalian stage 21 embryos

	Man	Baboon	Rhesus monkey	Mouse
Age (days)	53–54	(41) 43±1 (45)	(39) 40–44 (45)	14
Length (mm)	22–24	18–21	18–22	11–12
Neck region becomes apparent	+	+	+	+
Protruding forehead	+	+	+	+
Eyelids cover one-fourth of the eyes	+	+	+	–
Hair follicles at eyebrows	?	+	+	?
Auricula covers half of external auditory meatus	+	+	+	+
Primordial helix and anthelix	+	+	+	+
Upper jaw extends much further than lower jaw	+	+	+	+
Flexures at elbows and knees	+	+	+	+
Hands and feet approach each other	+	+	+	+
First sign of wrists	+	+	+	+
Distal phalanges of fingers separated	+	+	+	+
Feet parallel to midsagittal plane	+	+	+	+
Tips of toes separated	+	+	+	+
Wide lentoretinal space	+	+	+	+
Definitive lens fibers present	+	+	+	+
Anterior chamber differentiating	+	+	+	+
Cornea consists of three layers	+	+	+	+
Optic nerve fibers cross in chiasma opticum	+	+	+	?
Scleral condensation present	+	+	+	+
Lateral palatine processes started rotation	+	+	+	+
Further differentiation of nasal cartilages	?	+	+	+
Vomeronasal organ present	+	+	–	+
Pars intermedia of adenohypophysis thinwalled	+	+	+	+
Pars distalis: trabeculae + secondary vesicles	+	+	+	+
Pars distalis: vascular component increased	+	+	+	?
Neurohypophysis: lumen continuous with 3rd ventricle	+	+	+	?

that the corresponding embryos in the rat are 16–16 1/2 days of age and 12.7–14.5 mm long (Christie 1964).

The separation of the toes is the most striking, externally visible process that starts during developmental stage 21 and continues during later stages. The same conclusion can be drawn from the observations made by Christie (1964) on the rat. Among the external features characteristic of stage 21 embryos, as listed in Table 10, there appears to be only one distinction between the primates and the mouse, namely the width of the eyelids. In the mouse and in the rat (Christie 1964) the outgrowth of the eyelids does not start until stage 21, whereas in the primates it has already started during stage 20. Additionally it has to be mentioned that in contradistinction to the primates both the mouse and the rat embryos of stage 21 show numerous trunk hair papillae.

The most distinctive external characteristics of stage 21 embryos are: the distal phalanges of the fingers are separated; the arms and legs are in a flexed position at the elbows and knees; the hands and feet of both sides are approaching each other; and the tips of the toes are separated.

As in stage 20 the only internal difference between the various species analyzed is the absence of the vomeronasal organ in the rhesus monkey. The most outstanding internal feature of stage 21 embryos is the position of the lateral palatine processes: posteriorly they hold a vertical position lateral to the tongue; their medial parts point medially underneath the tongue and anteriorly they are apposed to the nasal septum and even fused with the median nasal process above the tongue. Similar observations were described by Christie (1964) for rat embryos, which in our opinion correspond to stage 21. Besides, the presence of a primordial anterior chamber and of definitive lens fibers in the eye and a progressive differentiation of the pars distalis of the adeno-hypophysis can also serve as criteria by which to define developmental stage 21.

3.10 Stage 22

3.10.1 Postconceptional Age and Length

The three rhesus monkey embryos representative of developmental stage 22 ranged in age from 44 ± 1 days to 48 ± 1 days p.c. The crown-rump length of the specimens varied between 20 and 25 mm.

3.10.2 External Features

In the embryos at stage 22 the external appearance is still dominated by the relatively large head (Fig. 20A). The elevation of the head has proceeded: the chin is detached from the ventral aspect of the chest. The outer form of the head is dominated by the progressive outgrowth of the forehead with the occiput lagging behind. The upper jaw still extends further than the lower jaw although the latter obviously has grown in comparison with the preceding stage. The eyes take an increasingly frontal position, whereas the eyelids now cover about one-third of the exterior of the eyes (Fig. 20A and C). The hair follicles at the eyebrows have extended over the root of the nose. The auricula has attained a rather definite form, due to the further differentiation of the helix and anthelix and also to the primordial tragus and antitragus (Fig. 20A).

The outgrowth of the extremities has proceeded, which is notably reflected in the lengthening of both the upper arms and upper legs. Compared with the preceding stage the flexures at both the elbows and the knees have increased (e.g., Figs. 20A, C and 17A, C). The wrists and the ankles are now also easily recognizable. The separation of the fingers has proceeded up to about the proximal phalanges (Fig. 20A and C). The fingers of the left and right hand partially overlap. The plantar surfaces of the feet still hold a position about parallel to the midsagittal plane with the tail between. The distal phalanges of the toes are separated; the toes of the left and right foot almost touch each other. Both the thumbs and the big toes hold an abducted position.

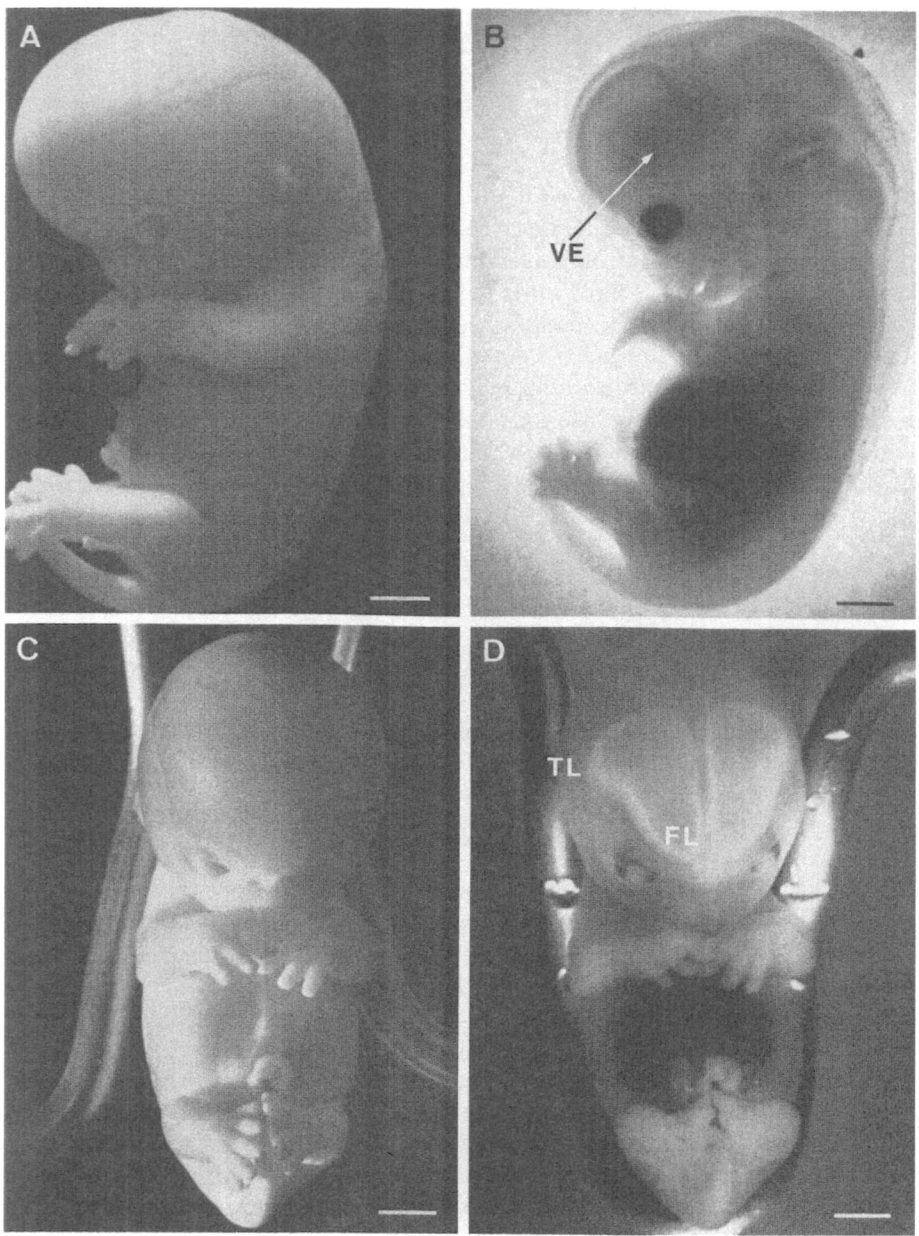

Fig. 20 A–D. Photomicrographs of a stage 22 embryo, aged 48 ± 1 days p.c. A and B. Lateral views, C and D. Ventral views; B and D cleared in methylbenzoate. Length of the *bars* 2 mm. Abbreviations: *FL*, frontal lobe; *TL*, temporal lobe; *VE*, ventricular eminence

3.10.3 Internal Features

The developmental state of the eyes of the embryos representative of stage 22 is clearly illustrated in Fig. 21A. Compared with the preceding stage (e.g., Fig. 18A and

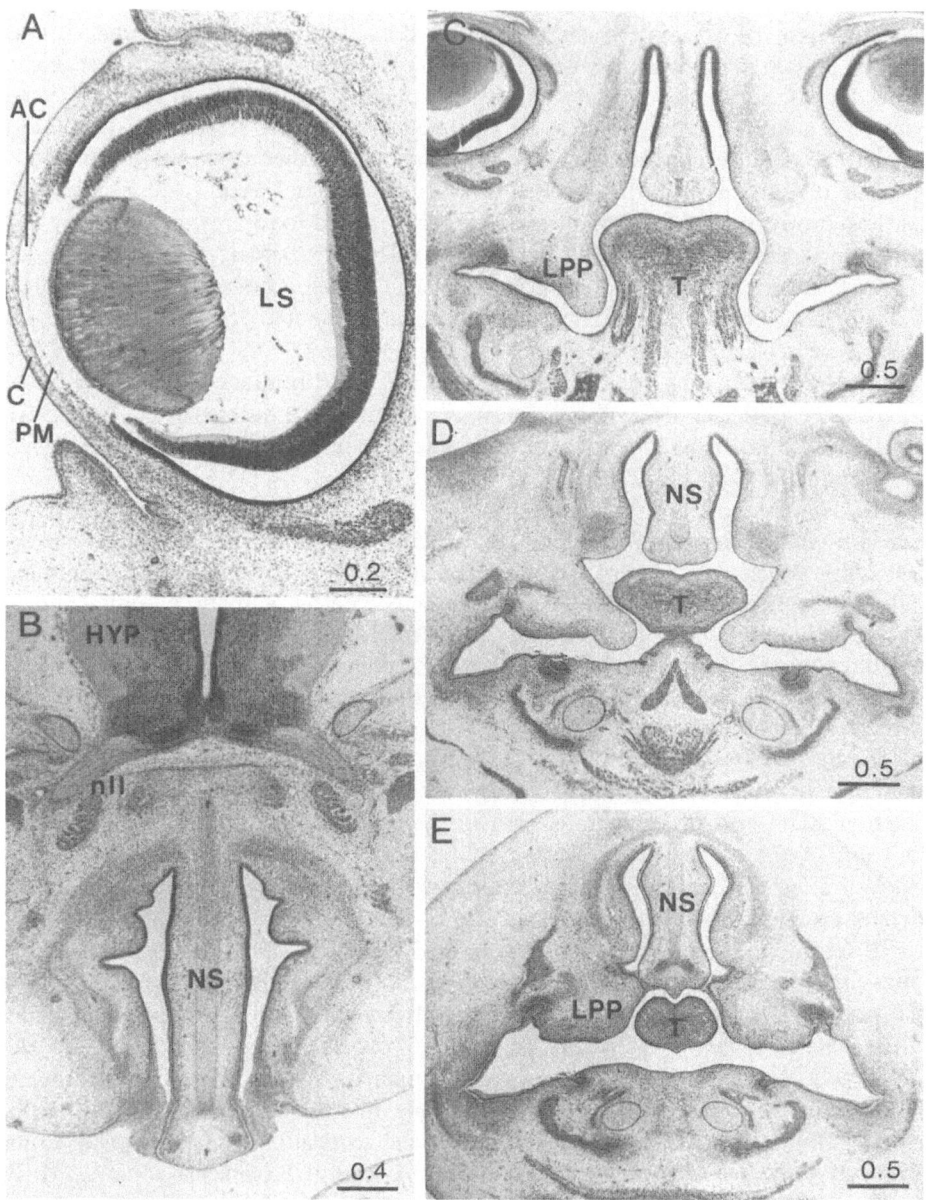

Fig. 21 A–E. Photomicrographs of stage 22 embryos. A and B. Frontal sections of an embryo aged 46 ± 1 days p.c., C, D and E. Horizontal sections of an embryo aged 48 ± 1 days p.c. Length of the *bars* noted in millimeters. Abbreviations: *AC*, anterior chamber; *C*, cornea; *HYP*, hypothalamus; *LPP*, lateral palatine process; *LS*, lentoretinal space; *NS*, nasal septum; *PM*, pupillary membrane; *T*, tongue; *n II*, optic nerve

B) the number of definitive lens fibers has considerably increased. In front of the lens the small anterior chamber is present, bordered posteriorly by the thin pupillary membrane and anteriorly by the cornea (Fig. 21A). The latter structure is composed of the epithelium and the mesothelium separated by the thickened substantia propria. The lentoretinal space has considerably widened and within it the primordial vitreous body is present. The constituent nuclear layer of the retina can be subdivided into a more densely packed outer layer and a less compact inner layer. In the optic nerve numerous collections of cell nuclei are scattered between the nerve fibers (Fig. 21B). The latter extend throughout the nerve and cross in the rather massive optic chiasma.

The oral and nasal cavities still show a broad communication (e.g., Fig. 21C and D). Posteriorly, the lateral palatine processes have started to rotate medially: they now hold a more or less oblique position (Fig. 21C). The intermediate parts of the palatine processes are oriented horizontally, pointing medially underneath the tongue (Fig. 21D). Anteriorly, however, the oral and nasal cavities are separated; the lateral palatine processes are fused with the nasal septum (Fig. 21E). Within the oral epithelium the dental lamina is differentiating. The septal and paraseptal cartilages are easily recognizable, whereas the cartilages of the alae majores have continued to differentiate. The conchae nasales project into the lumen of the nasal cavity, while the external nares are closed by an epithelial plug (Fig. 21B).

In comparison with the preceding stage the total volume of the hypophysis has somewhat increased. Within the pars distalis of the adenohypophysis numerous secondary vesicles are present, having a central lumen (Fig. 22A and B). Also many cells of the angiogenic tissue are arranged into rosettes without a central lumen. The pars infundibularis and the pars intermedia have hardly changed. Within the neurohypophysis a small lumen is still present within its intermediate part (Fig. 22A). In contradistinction to the previous stage, however, this lumen is no longer continuous with the third ventricle.

In the pictures of the embryos cleared in methylbenzoate (Fig. 20B and D) the extension of both the frontal and temporal lobes of the telencephalon is clearly illustrated and also the first outgrowth in the occipital direction can be recognized. In front of the cornu inferius of the lateral ventricle the ventricular eminence can be seen (Fig. 20B). Compared with the preceding stage the mesencephalon seems to have attained a more caudal position.

From the microscopic sections of the CNS it can be seen that caudally the ventricular eminence is one undivided structure (Fig. 22D). In its intermediate part (Fig. 22E) the ventricular eminence exhibits a shallow sulcus pointing to its original components — the medial and lateral ventricular ridges. More rostrally these two structures are even completely separated. The narrowing of the foramen of Monro has proceeded due to the outgrowth of the ventricular eminence and of the subjacent fiber mass of the capsula interna (Fig. 22D and E). The superficial cortical layer consists of up to ten layers of cells. The developing olfactory bulb is easily recognizable (Fig. 22D). Its lumen, the olfactory ventricle, is still continuous with the frontal part of the lateral ventricle.

Fig. 22 A–E. Photomicrographs of stage 22 embryos. A, C and D. Frontal sections of an embryo aged 46 ± 1 days p.c., B and E. Horizontal sections of an embryo aged 48 ± 1 days p.c. Length of the *bars* noted in millimeters. Abbreviations: *ADd*, adenohypophysis pars distalis; *Ad im*, adenohypophysis pars intermedia; *DI*, diencephalon; *DT*, dorsal thalamus; *E*, epiphysis; *HYP*, hypothalamus; *IC*, internal capsule; *NEU*, neurohypophysis; *OB*, olfactory bulb; *PC*, primary cortex; *VE*, ventricular eminence

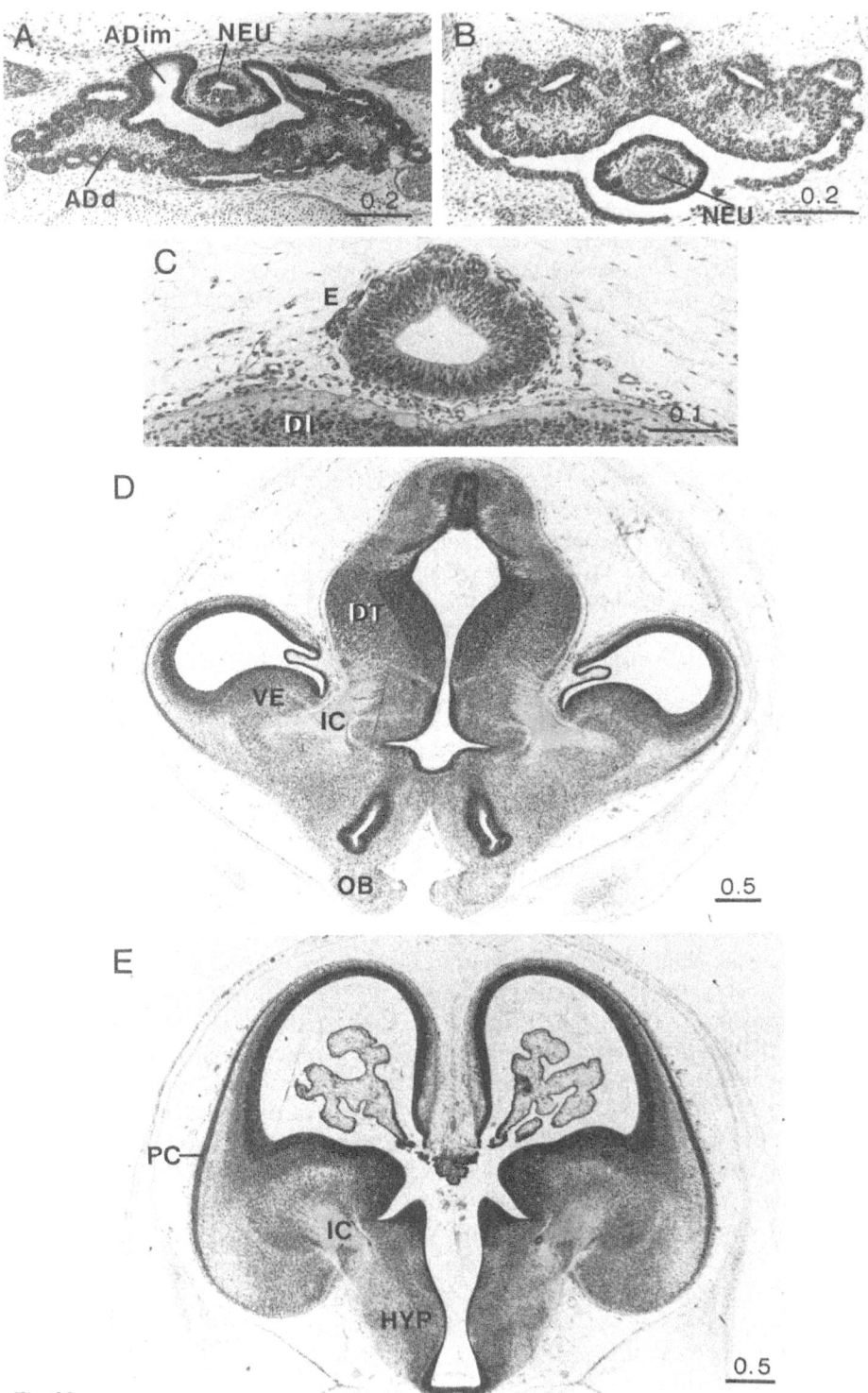

Fig. 22

In the diencephalon the third ventricle is no longer continuous with the lumen of the neurohypophysis. As a result the infundibular recess is formed. The progressive thickening of the thalamic wall caused a further narrowing of the adjacent part of third ventricle (Fig. 22D). The diencephalic roof gradually attains the character of a plexus choroideus (Fig. 22E). More caudally the epiphysis is situated, the lumen of which is now more triangular in cross section (Fig. 22C). Many mitoses can be observed at its ventricular surface. Moreover, the pineal gland obviously has started to differentiate: numerous mammilliform projections without a central cavity can be recognized; their cells are arranged into rosettes. Caudal to the epiphysis the broad commissura posterior marks the di-mesencephalic boundary.

The parts of the CNS caudal to the diencephalon develop in such a gradual way that a description of their microscopic conditions is not within the scope of this present study.

3.10.4 Comparison with Other Mammals

The data relating to those mammalian species on which a comprehensive description of stage 22 embryos is available in the literature are summarized in Table 11. Unfortunately, neigher rodents nor the marmoset could be included since in these species no data are available on embryos at this stage. Also, from the information given by Christie (1964) on the rat in can be concluded that in this species stage 22 embryos probably are about 16 1/2–17 days of age and about 14.5–16 mm long. Moreover, Butler (1974) mentioned a galago stage 22 embryo of unknown age, which had an overall length of 12.0 mm.

When the ages and lengths of the stage 22 embryos given in Table 11 are compared with those of stage 21 embryos (see Table 10) it becomes evident that the respective differences have remained stable: the human embryos being about 10 days older and 2 mm longer than those of the baboon and the rhesus monkey.

A comparison of the external characteristics of stage 22 embryos with those in previous stages reveals that no notable developmental processes start during this stage but those that started earlier are continued.

The most salient external features of the embryos representative of developmental stage 22 are: the chin is detached from the chest; all constitutive parts of the mature auricula are recognizable; the wrists and ankles are present; the fingers are separated up to the proximal phalanges; and the distal phalanges of the toes are separated.

From the internal characteristics listed in Table 11 it can be seen that the vomeronasal organ which in the rhesus monkey had already disappeared in stage 21 is now also absent in the baboon, whereas it is still recognizable in the human embryos.

The most discriminative internal qualities of stage 22 embryos are located within the oral and nasal cavities, which still exhibit a wide communication. The lateral palatine processes, however, are rotated over their full length: anteriorly, they are oriented horizontally above the tongue close to or even fused with the nasal septum; intermediately they are oriented horizontally underneath the tongue, and posteriorly they hold an oblique position lateral to the tongue. The start of the fusion of the lateral palatine processes was also reported by Christie (1964) in the representative rat embryos; moreover the nasal conchae are also recognizable during stage 22. Additional internal criteria of stage 22 embryos are: numerous lens fibers and a definitive anterior chamber are present in the eye; angiogenic tissue of the adenohypophysis

Table 11. Characteristics of primate stage 22 embryos

	Man	Baboon	Rhesus monkey
Age (days)	54–56	(44) 45±1 (47)	(43) 44–48 (49)
Length (mm)	23–28	21–23	20–25
Chin detached from chest	+	+	+
Upper jaw extends a little further than lower jaw	+	+	+
Eyelids cover approx. one-third of eyes	+	+	+
Definite helix and anthelix	+	+	+
Primordial tragus and antitragus	+	+	+
Flexures at elbows and knees deepened	+	+	+
Wrists and ankles present	+	+	+
Fingers separated up to proximal phalanges	+	+	+
Fingers of right and left hand partially overlap	+	+	+
Plantar surfaces of feet parallel to midsagittal plane	+	+	+
Distal phalanges of toes separated	+	+	+
Toes of right and left foot almost touch	+	+	+
Numerous lens fibers present	+	+	+
Definitive anterior chamber	+	+	+
Primordial vitreous body present	+	+	+
Numerous cell nuclei in optic nerve	+	+	+
Nuclear layer of retina subdivided into two layers	+	±	+
Oral and nasal cavities have wide communication	+	+	+
Lateral palatine processes: rotated over their full length	+	+	+
Started fusion of lateral palatine processes	–	–	+
Paraseptal cartilages evident	?	+	+
Vomeronasal organ present	+	–	–
Nasal conchae recognizable	+	+	+
External nares closed with epithelial plugs	?	+	+
Adenohypophysis: pars distalis exhibits rosettes	+	+	+
Neurohypophysis: lumen discontinuous of 3rd ventricle	+	+	+

is arranged in rosettes; and the lumen of the neurohypophysis is closed off from the third ventricle.

3.11 Stage 23

3.11.1 Postconceptional Age and Length

The age of the six embryos examined that were assigned to developmental stage 23 varied between 46 ± 1 days and 50 ± 1 days p.c. Their crown-rump length ranged from 24 to 30 mm.

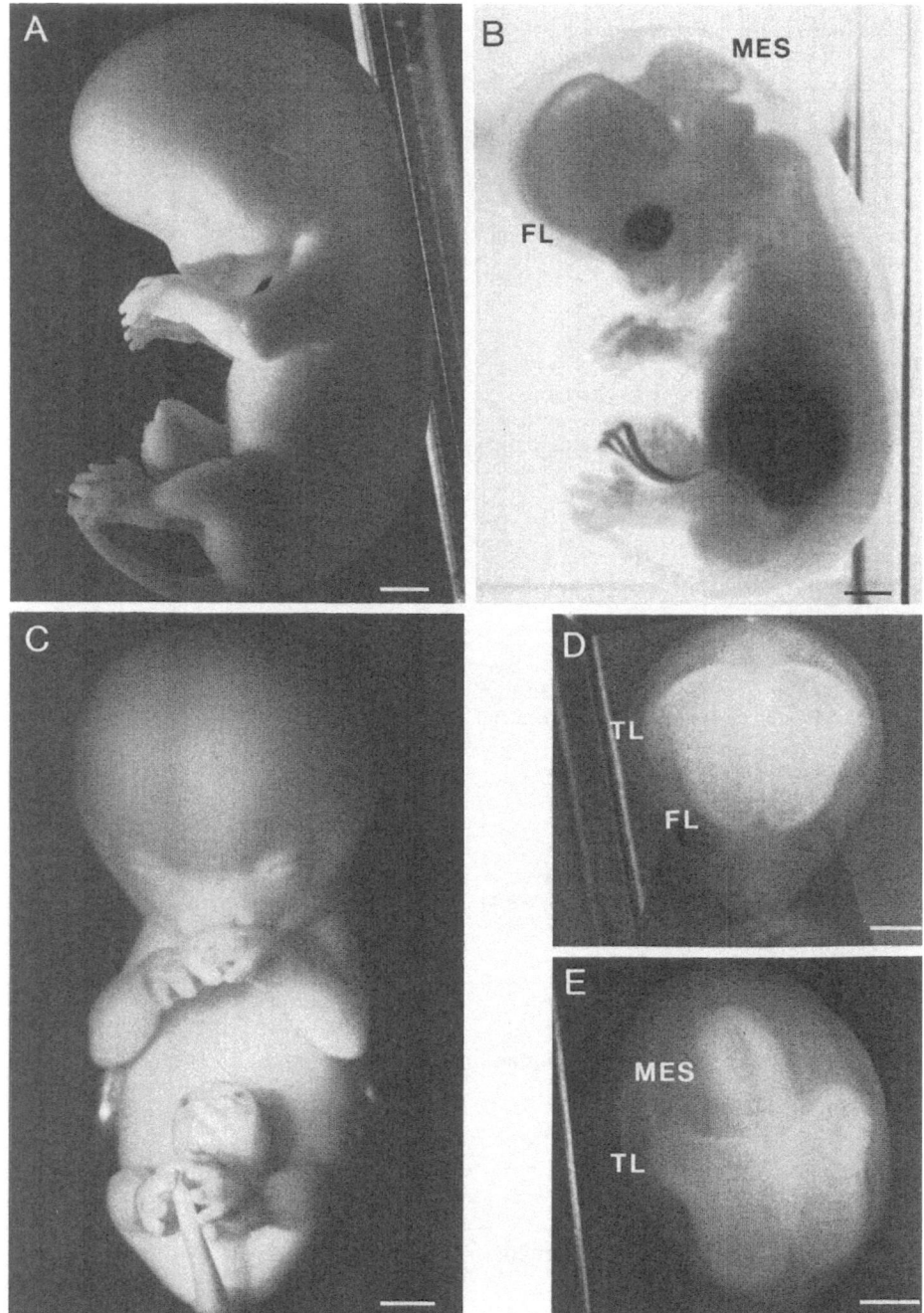

Fig. 23 A–E. Photomicrographs of a stage 23 embryo, aged 48 ± 1 days p.c. A and C. Lateral and ventral views, B, D and E. Lateral ventral and superior views of the embryo cleared in methyl-benzoate. Length of the *bars* 2 mm. Abbreviations: *FL*, frontal lobe; *MES*, mesencephalon; *TL*, temporal lobe

3.11.2 External Features

During stage 23 the progressive extension of the neck causes a further elevation of the head and thus of the chin (Fig. 23A and B). The volume of the neurocranium still exceeds that of the splanchnocranium, although the latter is gradually increasing. This phenomenon is also reflected in the fact that the lower jaw now protrudes almost as far as the upper jaw. The outer form of the head is less dominated by the protrusion of the forehead because of the gradual development of the occiput. In frontal pictures (Fig. 23C and D) the face is approximately wedge-shaped.

A characteristic feature of stage 23 is the closure of the eyelids, which is completed in the oldest members of the group (Fig. 23A and C); in the younger specimens the upper and lower eyelids have only started to fuse at their lateral and medial angles. The external ears assume their definitive shape (Fig. 23A), mainly because of the presence of the tragus and antitragus and the development of the tuberculum auriculae.

The extremities have increased considerably in length; the flexures at the elbows and knees too are more pronounced than in stage 22 (e.g., Figs. 23A, C and 20A, C). The forearms are raised along with the head, thereby keeping the position of the hands in front of the snout. The palmar surfaces of the hands face caudally and the left and right hand partially overlap (Fig. 23A and C). During stage 23 the separation of the proximal phalanges of the fingers occurs; thus the separation of the fingers is complete at the end of stage 23.

The position of the feet is similar to that in the preceding stage: they are very close together, the tail being between them, their plantar surfaces facing each other (Fig. 23C). In some specimens the toes of the left and right foot even touch. In the oldest members of the group the plantar surfaces of the feet start to turn caudally, beginning at the heels. During stage 23 the separation of the proximal phalanges of the toes occurs, so that at the end of the stage the toes are also completely separated. The distal phalanges of both the fingers and the toes have a swollen appearance.

3.11.3 Internal Features

The fusion of the upper and lower eyelids, which started at the lateral and medial angles of the eye, has proceeded toward the anterior pole of the eye. In the oldest specimens of stage 23 the fusion is even completed as is illustrated in Fig. 24A and B, where it can also be seen that the sclera is continuous with the substantia propria of the cornea. The presence of the vitreous body is evident (Fig. 24A–C). The nuclear layer of the retina consists of a more compact external neuroblastic layer and a less densely packed internal neuroblastic layer. The nerve fiber layer of the retina thickens toward its junction with the optic nerve (Fig. 24C). At this junction the hyaloid vessels can be seen entering the optic nerve at the center of the optic disc. Numerous clusters of cell nuclei are scattered throughout the optic nerve fibers. A nerve sheath can be clearly recognized around the optic nerve (Fig. 24C).

A characteristic internal feature of developmental stage 23 is the closure of the secondary palate: the lateral palatine processes fuse both with the nasal septum and with each other. At the end of stage 23 this process is completed and thus at that time the oral and nasal cavities are separated. This latter situation is illustrated in

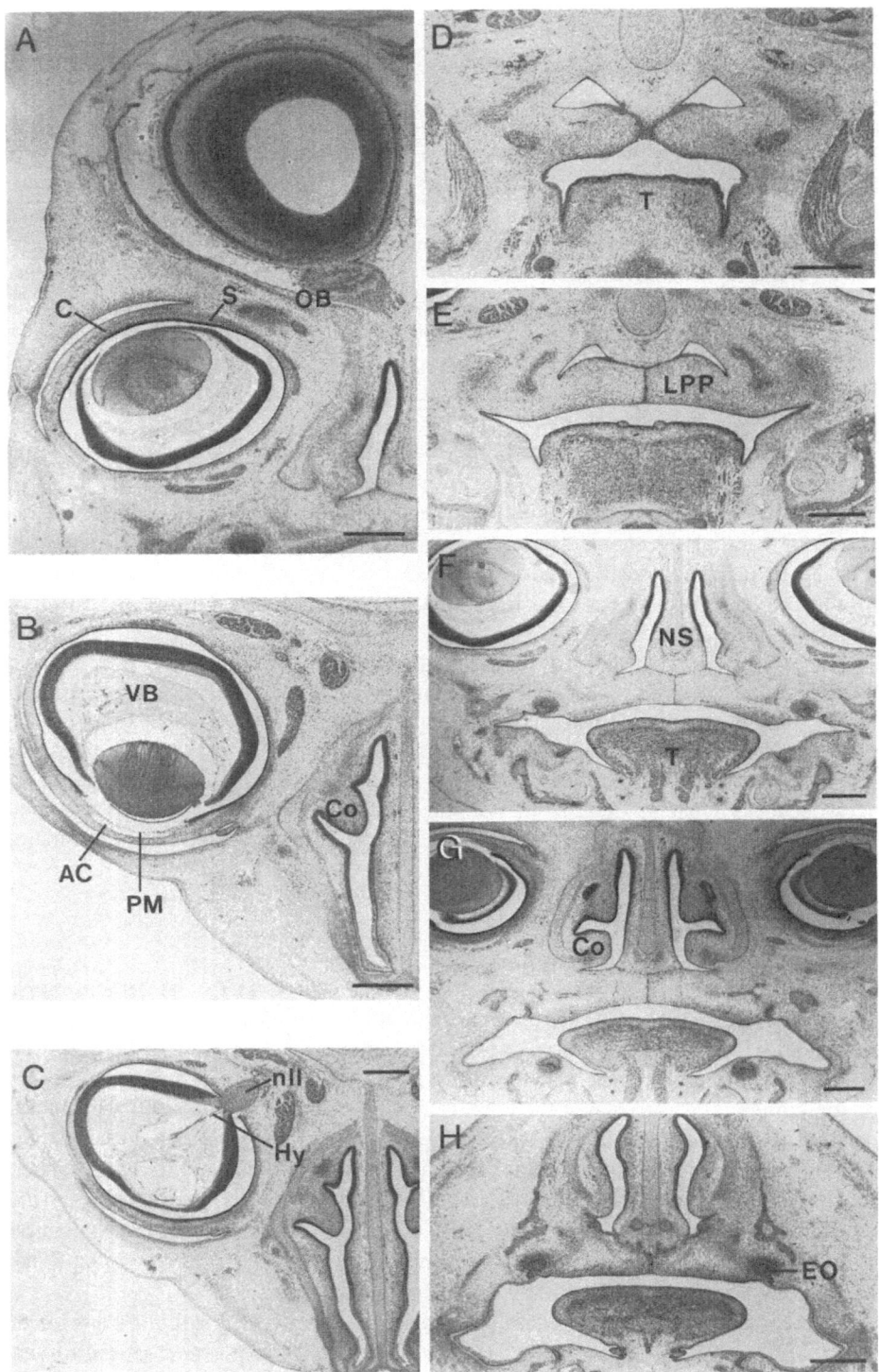

Fig. 24A–H

Fig. 24 (D–H), in which five transverse sections of the palatum of the same embryo are presented. Posteriorly (Fig. 24D), the adhesion of the epithelia of the nasal septum and the lateral palatine processes is clearly visible. Anteriorly (Fig. 24G and H), these epithelia have started to degenerate probably because of autolysis of their constituent cells as was suggested by Greene and Pratt (1976). The septal and paraseptal cartilages are well developed, as are the nasal conchae (Fig. 24B, C and G). The external nares are still closed by epithelial plugs. Local thickenings of the dental laminae represent the future enamel organs (Fig. 24F–H).

The total volume of the hypophysis has markedly increased compared with the preceding stage. Within the pars distalis of the adenohypophysis only a few secondary vesicles with a central lumen are still present: most of the cells are arranged in rosettes (Fig. 25A). The lumen of the neurohypophysis is almost completely obliterated, whereas its wall exhibits a slightly lobulated appearance in its distal part.

In the CNS the expansion of the telencephalon has proceeded markedly, as can be seen in the pictures of the methylbenzoate-cleared embryos (Fig. 23B, D and E). The frontal, temporal, and occipital lobes have extended in such a way that the diencephalon is now almost completely covered by the telencephalic hemispheres (e.g., Figs. 23B and 20B). Within the cerebral hemispheres the ventricular eminence is still faintly recognizable from the outside (Fig. 23B).

From the microscopic sections it can be seen that the merging of the two ventricular ridges into the single ventricular eminence has proceeded frontally (Fig. 25C and D). The cerebral stem area is to a large extent occupied by the capsula interna. The primary cortex has both widened and extended dorsally and frontally. The lumen of the olfactory ventricle has reduced but is still continuous with the lateral ventricle. The olfactory bulb itself has started to differentiate (Fig. 24A).

In the diencephalon the walls have progressively thickened, thereby reducing the lumen of the third ventricle to a large extent (Fig. 25C and E). In the thalamus the walls of both sides have even met in the midline forming a small adhesio interthalamica (massa intermedia). The roof of the third ventricle has definitively attained the character of a plexus choroideus. In its frontal part the epiphysis shows the compact aspect of a gland in which vesicles and follicles are present at its periphery (Fig. 25B).

In the mesencephalon the thickening of its wall has gradually proceeded up to its tectal part, but the mesencephalon still shows a wide ventricle. The intraventricular portions of the rhombic lips project far into the lumen of the fourth ventricle (Fig. 25F), with the result that the lumen of the lateral recesses is considerably reduced and is to a large extent filled with the plexus choroideus of the fourth ventricle. The brain stem area is a massive structure now in which various nuclei have started to differentiate (Fig. 25F).

Fig. 24 A–H. Photomicrographs of stage 23 embryos. A and D–H. Horizontal sections of an embryo aged 46 ± 1 days p.c., B and C. Frontal sections of an embryo aged 48 ± 1 days p.c. Length of the *bars* 0.5 mm. Abbreviations: *AC*, anterior chamber; *C*, cornea; *Co*, concha nasalis; *EO*, enamel organ, *Hy*, hyaloid vessels; *LPP*, lateral palatine process; *NS*, nasal septum; *OB*, olfactory bulb; *PM*, pupillary membrane; *S*, sclera; *T*, tongue; *VB*, vitreous body; *nII*, optic nerve

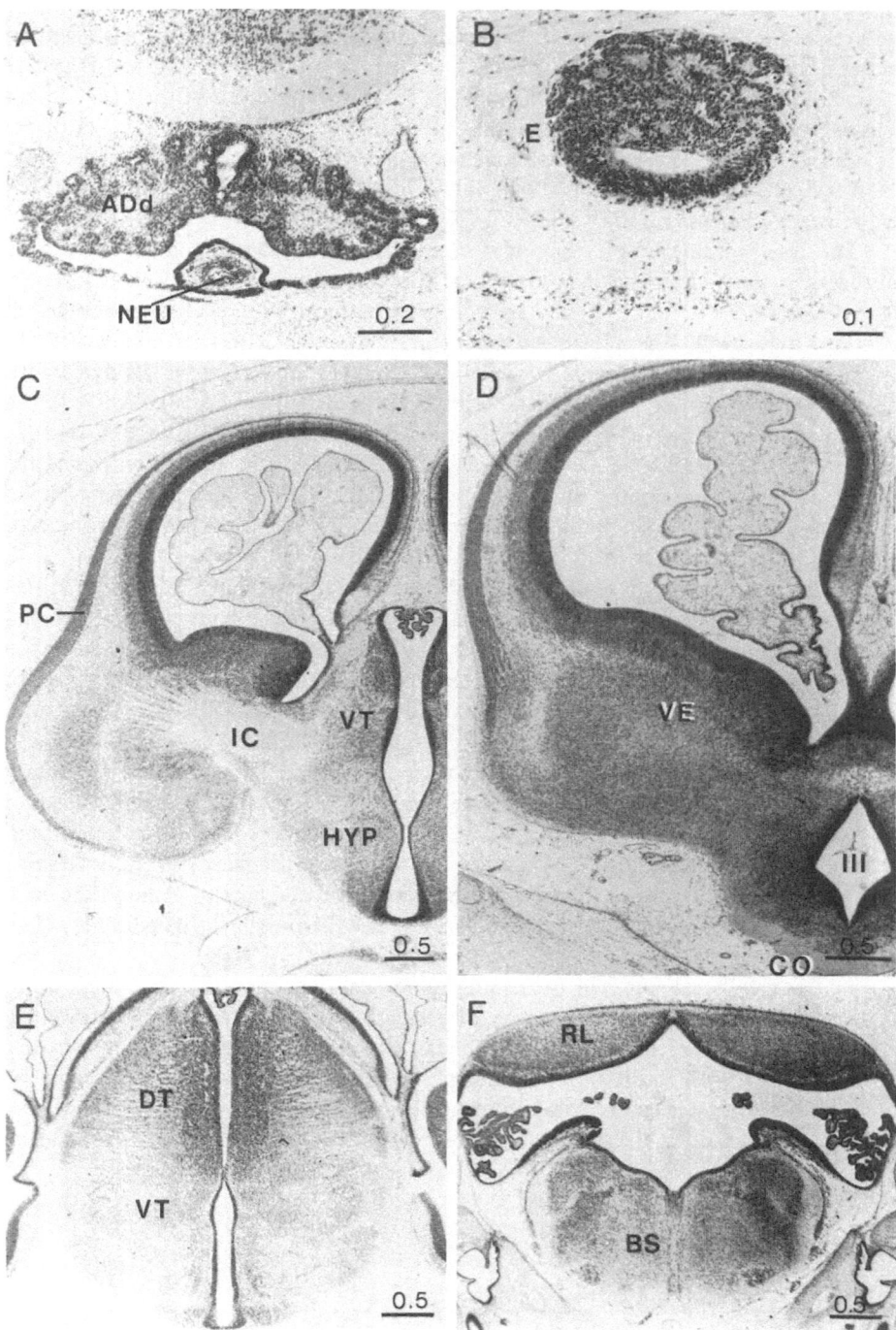

Fig. 25A–F

Table 12. Characteristics of primate stage 23 embryos

	Man	Baboon	Rhesus monkey
Age (days)	56−60	(45) 47±1 (50)	(45) 46−50 (51)
Length (mm)	27−31	25−28	24−30
Upper and lower jaw extend the same	+	+	+
Face is wedge-shaped	+	+	+
Developing occiput	+	+	+
Closure of the eyelids	±	±	±
Definitive auricula	+	+	+
Deep flexures at elbows and knees	+	+	+
Flexures at wrists and ankles	+	+	+
Forearms raised	+	+	+
Separation of proximal phalanges of fingers	+	+	+
Separation of proximal phalanges of toes	+	+	+
Plantar surfaces of feet turn caudally	±	±	±
Toes of both feet touch	+	+	+
Distal phalanges of fingers and toes swollen	+	+	+
Fusion of upper and lower eyelids	±	±	±
Definitive lens structure	+	+	+
Wide cornea present	+	+	+
Vitreous body present	+	+	+
Nerve sheath around optic nerve	+	+	+
Closure of secondary palate	+	+	+
Prominent conchae nasales	+	+	+
Dental lamina: future enamel organs	?	+	+
Volume hypophysis markedly increased	+	+	+
Pars distalis of adenohypophysis: most cells arranged in rosettes	+	+	+
Neurohypophysis: lumen almost obliterated	+	+	+
Distal wall of neurohypophysis: lobulated	+	+	+

3.11.4 Comparison with Other Mammals

It has already been stressed that developmental stage 23 is defined as the last stage of the embryonic period at the end of which the secondary palate is closed. The end of stage 19 in the Chinese hamster in which the closure of the secondary palate occurs as described by ten Donkelaar et al. (1979), therefore must be considered comparable to the end of primate stage 23. The same holds for Theiler's (1972) stage 23 in the mouse. Nevertheless, a large number of the external and internal features characteristic of primate stage 23, as listed in Table 12, were not described in detail by the authors mentioned; hence these two rodent species could not be inserted in this table.

Fig. 25 A−F. Photomicrographs of stage 23 embryos. A−E. Horizontal sections of an embryo aged 46 ± 1 days p.c., F. Frontal section of an embryo aged 48 ± 1 days p.c. Length of the *bars* noted in millimeters. Abbreviations: *ADd*, adenohypophysis pars distalis; *BS*, brain stem; *CO*, optic chiasm; *DT*, dorsal thalamus; *E*, epiphysis; *HYP*, hypothalamus; *IC*, internal capsule; *NEU*, neurohypophysis; *PC*, primary cortex; *RL*, rhombic lip; *VE*, ventricular eminence; *VT*, ventral thalamus; *III*, third ventricle

To the figures about the ages and lengths of the primates presented in Table 12 it should be added that in the marmoset stage 23 occurs at about 83 days (Phillips 1976), whereas galago embryos representative of stage 23, which were of unknown age, measured about 14.0 mm (Butler 1972). In the rodents the following data on age and length of embryos comparable with stage 23 were presented by the respective authors; in the Chinese hamster 15 3/4–16 days and 10.5–11 mm; in the mouse 15 days and 12–14 mm; whereas comparable rat embryos are aged 17–17 1/2 days and are 16–17.6 mm long (Christie 1964).

The most prominent external feature of primate stage 23 embryos is that during this stage the separation of both the fingers and the toes is completed. The same holds for the mouse stage 23 embryos described by Theiler (1972). In the comparable stage 19 (ten Donkelaar et al. 1979) embryos of the Chinese hamster the toes are separated in the forelimb but not yet in the hindlimb. Similar results were obtained by Christie (1964) in the rat: in his stage 30 the forepaw digits are completely separated, whereas the hindpaw digits are separated during stage 31 (17 1/2–18 days; 17.6–19.1mm). In the Chinese hamster the separation of the hindlimb toes occurs during the subsequent 'fetal stage 1' as defined by the authors.

Another external feature characteristic of primate stage 23 embryos is the progressive closure of the eyelids. In most specimens the eyelids are only left open at the anterior poles of the eyes; in the oldest representatives the eyes were even completely closed in both the baboon and the rhesus monkey. In the rodents the closure of the eyelids clearly lags behind: in all rodent species analyzed the eyes were reported open during stage 23. In the rat the eyes close during Christie's (1964) stage 32, being 18–18 1/2 days old and 19.1–22 mm long. In the Chinese hamster closure of the eyelids was reported (ten Donkelaar et al. 1979) at the age of 18 days.

As regards the internal characteristics the closure of the secondary palate by definition is the most salient feature of the last stage of the embryonic period. Therefore this criterion is represented in all primate stage 23 embryos analyzed, including the marmoset and the galago. In the Chinese hamster this feature was reported during stage 19 (ten Donkelaar et al. 1979), in the mouse during stage 23 (Theiler 1972), and in the rat during stage 30 (Christie 1964). A comparison of the other internal features listed in Table 12 with those in rodents is not possible since the relevant details were not available.

4 Discussion

As was described in the Introduction to this review, Streeter (1942, 1945, 1948, 1951) subdivided the human embryonic period into 23 successive developmental stages on the basis of internal and external characteristics. Later these stages were further elaborated by O'Rahilly and his colleagues (1973–1979). The studies of Hendrickx (1971) on the baboon and Phillips (1976) on the marmoset rendered it likely that all primate species follow the same pattern of 23 consecutive developmental stages. The extensive studies of Theiler (1972) on the mouse and ten Donkelaar et al. (1979) on the Chinese hamster made it reasonable to assume that rodents also show a developmental pattern that is essentially similar to that in primates. Especially in the earlier stages rodent embryos concurred fairly well with those of primates. In the present investigation an attempt has been made to adduce arguments in support of a universal staging system of the embryonic development in all mammals.

From the data presented in the preceding section it appears that the embryonic period of all mammalian species analyzed hitherto can be subdivided into 23 successive stages, which can be arranged into three groups representing consecutive phases of embryonic development. The first phase is composed of the presomite stages 1–8; the second phase covers the somite stages 9–12, and the third phase, also called the organogenetic period, comprises the postsomite stages 13–2.

During the presomite stages the fertilized egg (stage 1) develops into the embryonic disc with a primitive streak and an incipient neural plate (stage 8). In view of the characterization of these stages it is unlikely that any new information could be added to the extensive descriptions and pictures presented by earlier authors (e.g., Heuser and Streeter 1941; Streeter 1942; Hertig et al. 1956; Heuser and Corner 1957; Hendrickx 1971; O'Rahilly 1973a).

During the somite stages the neural plate stage with 0–3 pairs of somites (stage 9) develops into the neural tube stage with 21–29 pairs of somites (stage 12). The intermediate stages 10 and 11 are accurately defined by the number of somites present: stage 10 has 4–12 pairs and stage 11 has 13–20 pairs of somites. For detailed information the reader is referred to the investigations of Heuser and Streeter (1941), Streeter (1942), Heuser and Corner (1957), Hendrickx (1971), Theiler (1972), Phillips (1976), and ten Donkelaar et al. (1979). In all species described, both primates and rodents, the anterior neuropore of the neural tube invariably closes during developmental stage 11. On the other hand the closure of the posterior neuropore occurs during stage 12 in primates and during stage 13 in rodents. Apart from this exception a high degree of conformity exists in the literature on the descriptions of the developmental stages 1 through 12 of different mammalian species.

During the postsomite or organogenetic developmental stages the mammalian embryo having more than 30 pairs of somites and a closing or closed neural tube (stage 13) develops into the embryo in which all organs and organ systems are present and the secondary palate is closed (stage 23). The transition from embryo into fetus, marking the end of stage 23, is defined as the time of closure of the secondary palate, as introduced by Hendrickx (1971) and recently adopted by O'Rahilly (1978) and ten Donkelaar et al. (1979). This point of transition is much more suitable than Streeter's (1951) criterion, i.e., the invasion of bone marrow into the cartilaginous precursor in the humerus.

From the data available in the literature on the organogenetic period in various mammalian species the tentative conclusion can be drawn that the sequence in which the individual organs are formed is basically similar in all mammals (see also Wilson 1973) but that the timing of the events occurring within particular organs is specific for closely related groups such as rodents or primates (see also ten Donkelaar et al. 1979). The present investigation was executed primarily to provide a detailed morphological description of the successive organogenetic stages in the rhesus monkey together with a reliable timetable of the events occurring during that period. Second, a more accurate comparison was made of the corresponding stages in different mammalian species to find out whether the thesis proposed could be substantiated or had to be modified.

Table 13. External and internal features of the organogenetic stages in mammals

External features		Internal features
Stage 13	Posterior neuropore closed Prominent 1st and 2nd branchial bars Definite forelimb bud	Straight olfactory placode Straight lens placode Three brain vesicles
Stage 14	Deep nasal pit Lens pit Definite hindlimb bud	Deep indented olfactory placode Deep indented lens placode Subdivision prosencephalon
Stage 15	Lens pit absent Vague retinal pigmentation Subdivision of forelimb bud	Closure of lens vesicle First retinal pigment granules Rhombomeres present
Stage 16	Auricular hillocks on hyoid bar Handplate present Subdivision hindlimb bud	Lens vesicle detached from ectoderm Numerous retinal pigment granules Distinct cerebral hemispheres
Stage 17	Auricular hillocks on mandibular bar First sign of fingerrays Footplate appearing	Lens cavity crescentic Medial + lateral nasal processes fused Neurohypophysis evaginated
Stage 18	Primitive nostrils Distinct fingerrays Definite footplate present	Lumen lens cavity slitlike Primary lens fibers present Onset of differentiation of cartilages
Stage 19	Auricular hillocks start fusion Rim of handplate crenated Distinct toerays present	Lens cavity obliterated Bucconasal membrane ruptured Processus palatini bilateral to the tongue
Stage 20	Definitive auricula recognizable Distal fingertips separated Rim of footplate crenated	Lens is a solid sphere Adenohypophysis starts differentiation Anterior parts of lateral palatine processes start rotation
Stage 21	Outgrowth eyelids started Distal phalanges of fingers separated Flexures at elbows and knees Tips of toes separated	Definitive lens fibers present Anterior chamber present Optic chiasm present Lateral palatine processes anteriorly start fusion; intermediate parts horizontally
Stage 22	Chin detached from chest Wrists and ankles present Middle phalanges of fingers separated Distal phalanges of toes separated	Numerous lens fibers present Lumen of neurohypophysis disconnected Nasal conchae recognizable Rotation of lateral palatine processes completed
Stage 23	Closure of eyelids started Separation of fingers and toes completed Flexures at wrists and ankles Hands rotated	Fusion of upper and lower eyelids started Nerve sheat around optic nerve Lumen of neurohypophysis becoming obliterated Closure of secondary palate

In Table 13 a survey is given of the most salient external and internal features characteristic of the developmental stages during the organogentic period. Regarding the primates it can be stated that all features listed apply to all species analyzed. As far as the rodents are concerned some minor deviations of the timing of some events have to be mentioned.

In the rodents the posterior neuropore closes during stage 13, whereas in primates it is closed during stage 12. The indentation of the lens placode seems to start somewhat earlier in the rodents (stage 13) than in the primates (stage 14); however, the closure of the lens vesicle both in rodents and in primates occurs during stage 15. Both the appearance and the subdivision of the forelimb buds and the hindlimb buds was recorded somewhat earlier in the rodents than in the primates.

The only other striking deviation found both between the different rodents and between rodents and primates was the time of closure of Rathke's pouch. In the Chinese hamster this closure was described as occurring during stage 14 (ten Donkelaar et al. 1979), whereas in the mouse Rathke's pouch is still open during stage 16 (Theiler 1972). In the primates the closure of Rathke's pouch was invariably recorded during stage 18. Much less disparity was found in the time of rupture of the bucconasal membrane: in the rodents it occurs during stage 18, in the primates in stage 19.

The outgrowth of the eyelids starts somewhat earlier in the primates (stage 20) than in the rodents (stage 21). The closure of the eyelids in the primates is completed at the end of stage 23 or soon after the onset of the fetal period. In the rodents the fusion of the eyelids invariably occurs during the early fetal stages.

The separation of the fingers both in primates and in rodents starts during stage 20 and is completed during stage 23. The separation of the toes starts in both groups of mammals during stage 21; however, this process is completed in primates during stage 23, whereas in rodents it ends at the beginning of the fetal period.

In noting these differences it must be emphasized that a much greater number of both external and internal features characteristic of the individual developmental stages concur fairly well in primates and rodents. Moreover it must be kept in mind that a more detailed analysis of the rodent stages 18 through 23 possibly might add still more data in support of the consistency between primates and rodents.

In the present investigation the actual postconceptional ages found for the rhesus monkey embryos representative of each of the stages 13 through 23 varied to some extent. Therefore, in Fig. 26 the range between the highest and the lowest age per developmental stage is illustrated. It should be noted that the inexactitude of the ages resulting from the mating procedure (±1 day) is not taken into consideration. From Fig. 26 it can be concluded, first, that each developmental stage covers a period of about four days. Second, the 4 day period of each developmental stage overlapped with that of the preceding stage by about 3 days. Or, in other words, at each postconceptional day of age about three possible developmental stages can be found. In this way it could be shown that in the rhesus monkey the postconceptional age is by no means a valid method for characterizing the embryos. A third conclusion that can be drawn from Fig. 26 is that the correlation between developmental stage and embryonic age in the rhesus monkey is nonlinear.

In Table 14 the data available on the age of the embryos representative of the consecutive organogenetic stages in various mammalian species are summarized. Greater variation, however, was observed in the baboon (Hendrickx 1971) and the rhesus monkey (present investigation) than is presented in the table. From Table 14 it might be concluded that apart from the rhesus monkey and the marmoset the ages at two successive stages do not overlap in other mammalian species. But it is a well-known fact that considerable variation in development can also be observed in rodents, not only between different litters of the same embryonic age but even between littermates (see also Ziehen 1906; Otis and Brent 1954; Christie 1964; Edwards 1968;

69

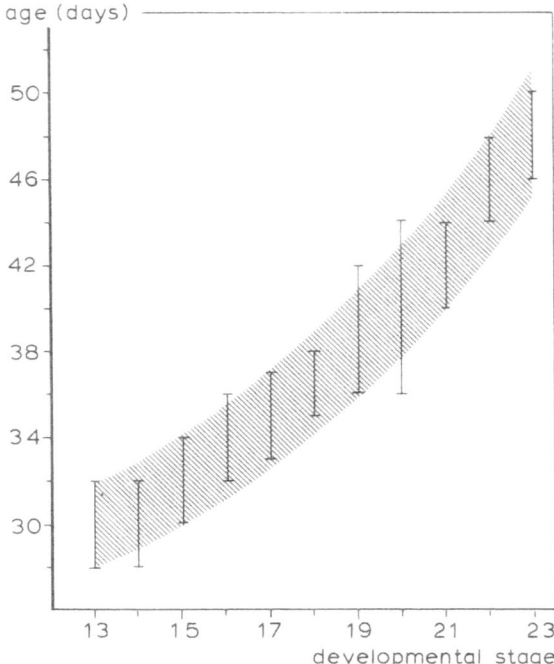

Fig. 26. Ranges of the ages of the rhesus monkey embryos representative of development stages 13–23

Table 14. Embryonic age (in days) of the organogenetic stages in various mammalian species

Develop- mental stage	Man[a]	Baboon[b]	Rhesus monkey[c]	Marmo- set[d]	Rat[e]	Chinese hamster[f]	Mouse[g]
13	28–32	28–30	28–30	61–67	$11^{1/2}$–12	$11^{1/2}$–12	10
14	31–35	29–31	30–32	63–?	12–13	12–$12^{1/2}$	$10^{1/2}$
15	35–38	30–32	30–32	61–73	13–$13^{3/4}$	$12^{1/2}$–13	11
16	37–42	32–34	32–34	66–83	$13^{3/4}$–14	$13^{1/2}$	$11^{1/2}$
17	42–44	34–36	34–36	61–77	14–$14^{1/2}$	?	12
18	44–48	36–38	35–38	?	$14^{1/4}$–$14^{3/4}$	14	?
19	48–51	38–40	36–42	±75	$14^{3/4}$–$15^{1/2}$	15–$15^{1/4}$	13
20	51–53	40–42	38–42	?	$15^{1/2}$–16	?	?
21	53–54	42–44	40–44	80	16–$16^{1/2}$	?	14
22	54–56	44–46	44–48	?	$16^{1/2}$–17	?	?
23	56–60	46–48	46–50	83	17–$17^{1/2}$	$15^{3/4}$–16	15

[a] O'Rahilly (1979); [b] Hendrickx (1971); [c] present study; [d] Phillips (1976); [e] Christie (1964); [f] ten Donkelaar et al. (1979); [g] Theiler (1972)

Theiler 1972; Festing 1976; ten Donkelaar et al. 1979). The figures given by the authors mentioned and summarized in Table 14 therefore cannot be absolute since more variation is very likely. It thus seems reasonable to assume a certain amount of overlap also in mammalian species other than the rhesus monkey.

Secondly, it might be suggested from Table 14 that in some mammalian species a linear correlation exists between the developmental stages and the respective ages

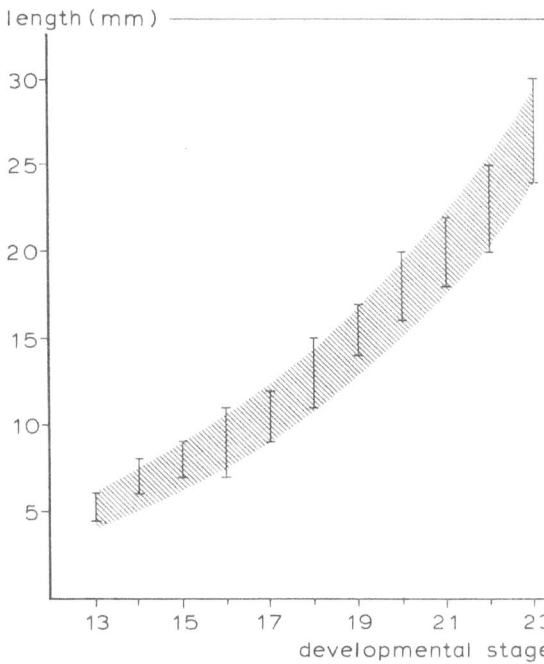

Fig. 27. Ranges of the lengths of the rhesus monkey embryos representative of developmental stages 13–23

of the embryos. In our opinion, however, this is not very likely, since at least in the nonhuman species the linear relationship probably results from the procedure followed in the analysis of the embryos. The most appropriate way of collecting specimens over the whole organogenetic period is to draw samples at equal time intervals. If, however, of the embryos examined at a specific time only the one being either the medium or the most advanced one in development is subsequently described as the representative of a stage, any variation is eliminated. This method, which was used, for example, by Theiler (1972) on the mouse, inevitably results in a linear relationship between developmental stages and age of the embryos. In our investigation the procedure after the collection of the embryos was the reverse: the developmental features of the various embryos were conceived of as the primary characteristics, while the ages were regarded only as additional data. In this way a more objective concept of the relationship between developmental stage and age of the embryos is achieved.

In Fig. 27 the length of the rhesus monkey embryos representative of the organogenetic developmental stages recorded in the present investigation is shown. It can be seen that an embryo of a specific length might belong to any of two or three different developmental stages. Thus it can be concluded that the length of the embryos, although a little more discriminative than the age of the embryos, also cannot serve as a criterion to define a developmental stage. Second, it might be concluded from Fig. 27 that the variation in length amounts to less in the earlier stages than in the later stages. Yet the percentage of deviation from the average length decreases from about 20% in the earlier stages to about 10% in the later stages. Third, it is clearly illustrated that the length of the embryos does not increase linearly with progressing developmental stages. Per developmental stage the growth gradually increases from about 2 mm in stage 14 to about 5 mm in stage 23.

71

Table 15. Embryonic length (in mm) of the organogenetic stages in various mammalian species

Develop-mental stage	Man[a]	Baboon[b]	Rhesus monkey[c]	Marmo-set[d]	Rat[e]	Chinese hamster[f]	Mouse[g]
13	4−6	4.5−6	4.5−6	2.4−4.8	3.0−4.1	2.8−4.2	3.1−3.9
14	5−7	6−7	6−8	4.1−?	4.1−5.8	4.6−5.1	3.5−4.9
15	7−9	6−8	7−9	6.8−7.8	5.8−7.9	5.1−6	5−6
16	8−11	7−9	7−11	8.3−9.2	7.9−9.4	5.6−6.3	6−7
17	11−14	10−13	9−12	±12.0	9.4−10.3	?	7−9
18	13−17	14−17	11−15	?	10.3−11.5	7.3−8.6	?
19	16−18	16−17	14−17	12.1−12.4	11.5−12.1	8.5−10	9−11
20	18−22	17−18	16−20	?	12.1−12.7	?	?
21	22−24	18−21	18−22	14.4−15.8	12.7−14.5	?	11−12
22	23−28	21−23	20−25	?	14.5−16	?	?
23	27−31	25−28	24−30	?	16−17.6	10.5−11	12−14

[a] O'Rahilly (1979); [b] Hendrickx (1971); [c] present study; [d] Phillips (1976); [e] Christie (1964); [f] ten Donkelaar et al. (1979); [g] Theiler (1972)

In Table 15 a survey is given of the lengths of the embryos representative of the successive organogenetic stages in various mammalian species, as reported by different authors. It must be stressed here, however, that the figures given in Table 15 can only be indicative since both the time of measurement of the embryos (before or after fixation) and the kind of fixation (determinant of the amount of shrinkage of the embryos) vary in the studies mentioned. From Table 15 it can be deduced that the conclusions described earlier for the rhesus monkey also apply to other primates, although in the marmoset the material is still rather incomplete. It also seems justifiable to draw the same conclusions for the rodents, especially the rat. For both the Chinese hamster and, although to a lesser extent, the mouse, however, more closely graded material is needed to verify the statements mentioned.

It must be emphasized again that in the present investigation substantial evidence is provided to support the thesis that neither the age nor the length of the embryos is a valid criterion to define a developmental stage. On the contrary, developmental characteristics are the only appropriate criteria upon which the staging of embryos can be based. This is true especially for studies in which the development of a particular organ is analyzed.

The chronology of the prenatal development in several mammalian species is summarized in Table 16. From this table the duration of the respective presomite phases (starting on embryonic day 1), somite phases, postsomite or organogenetic phases, and the fetal periods can easily be deduced. The presomite phase in the rodents lasts 8−10 days and in the primates 19−23 days. Only the marmoset clearly shows a protracted presomite phase, which according to Phillips (1976) has to be attributed to a slower rate of development rather than to a delayed implantation. Whether this hypothesis is correct remains questionable since the marmoset does not show any retardation of its development during later phases.

The somite phase in the rodents takes about 2 days, whereas in the primates it lasts about 8−10 days. The organogenetic phases range from about 4−6 days in the rodents to about 20 days in the subhuman primates and to about 30 days in man.

Table 16. Chronology of the prenatal development in various mammalian species

	Time of implantation	Start of somite phase	Start of organogenetic phase	Start of fetal period	Time of delivery
Man[a]	E_{5-6}	E_{19-21}	E_{28-32}	E_{56-60}	$E_{260-280}$
Baboon[b]	$\pm E_9$	$\pm E_{23}$	E_{28-30}	E_{46-48}	$E_{172-178}$
Rhesus monkey[c,d]	$\pm E_9$	$\pm E_{20-21}$	E_{28-30}	E_{46-50}	$E_{164-170}$
Marmoset[e]	unknown	$\pm E_{55-59}$	E_{61-67}	$\pm E_{83}$	$E_{145-155}$
Rat[f]	$E_{5.5-6}$	$E_{9.5-10}$	$E_{11.5-12}$	$E_{17-17.5}$	E_{21-22}
Chinese hamster[g]	$E_{5.5-6}$	E_{10}	$E_{11.5-12}$	E_{16}	E_{20-21}
Mouse[h]	$E_{4.5}$	E_8	E_{10}	E_{15}	E_{15}

[a] O'Rahilly (1979); [b] Hendrickx (1971); [c] Heuser and Streeter (1941); [d] Present investigation;
[e] Phillips (1976); [f] Christie (1964); [g] ten Donkelaar et al. (1979); [h] Theiler (1972)

The most striking differences appear in the duration of the respective fetal periods. In the rodent group the fetal period is rather uniform in lasting about 4–5 days; however, in the primate group the fetal period varies from about 65 days in the marmoset to about 120–130 days in the rhesus monkey and the baboon, respectively, and to about 210 days in man. This implies that the ratio between embryonic and fetal periods is quite different in rodents from that in primates, but also varies considerably among the primates themselves. The ratio between embryonic and fetal periods is about 3:1 in the rodents, about 1:1 in the marmoset, about 1:2.5 in the rhesus monkey and the baboon, and about 1:3.5 in man. These figures prove that designations like 'first half' or 'second half' of the prenatal period, which are often employed in comparative studies, are meaningless. Thus a comparison between rodent, subhuman primate, and human prenatal development can only be correctly made when corresponding developmental periods are analyzed.

From the ratios between the respective embryonic and fetal periods the assumption could be made that the length of the fetal period might be indicative of the phylogenetic position of each mammalian species. Nevertheless, extensive research on the fetal period in various mammalian species has to be accomplished before any substantive evidence in support of this hypothesis can be adduced.

Summarizing the results on the staging of the embryos both in primates and in rodents two main conclusions can be drawn. First, both groups of mammals show a common basic developmental plan in which the organogenetic sequence of particular organs is highly similar. Second, during the organogenetic period the developmental state of the individual organs appears to be highly correlated. This implies that both in primates and in rodents each organogenetic developmental stage is characterized by a combination of specific developmental characteristics of various organs, as is summarized in Table 17. Moreover it has to be mentioned that, irrespective of the species-specific anatomic differences, the subsequent developmental stages apparently proceed according to a species-specific time schedule, which was given in Table 14.

Table 17. Survey of the main organogenetic features of the prenatal stage in primates and rodents

5 Summary

In the present investigation the successive organogenetic stages 13 through 23 in the rhesus monkey (*Macaca mulatta*) are described on the basis of external and internal characteristics of the representative embryos in accordance with Streeter's classification. For each stage a comparison is made with other primates, including man, and with rodents. This comparison revealed (1), that a common basic developmental plan exists in all mammalian species studied so far, i.e., the sequence in which the individual organs are formed is highly similar, and (2) that there is a correlation between the developmental states of the organs. From these results a common staging system was deduced applying to all primate and rodent species analyzed.

References

Asling CW, van Wagenen G (1967) A note on the development of the secondary palate in the rhesus monkey (*Macaca mulatta*). Arch Oral Biol 12:909–910

Bollert JA, Hendrickx AG (1971) Morphogenesis of the palate in the baboon (*Papio cynocephalus*). Teratology 4:343–354

Boyd JD (1956) Observations on the human pharyngeal hypophysis. J Endocrinol 14:66–77

Burdi AR, Faist K (1967) Morphogenesis of the palate in normal human embryos with special emphasis on the mechanisms involved. Am J Anat 120:149–160

Butler H (1972) The chronology of embryogenesis in the lesser galago: a preliminary account. Folia Primatol (Basel) 18:368–378

Christie GA (1964) Developmental stages in somite and post-somite rat embryos, based on external appearance, and including some features of the macroscopic development of the oral cavity. J Morphol 114:263–268

Coleman RD (1965) Development of the rat palate. Anat Rec 151:107–118

Donkelaar HJ ten, Geijsberts LGM, Dederen PJW (1979) Stages in the prenatal development of the Chinese hamster (*Cricetulus griseus*). Anat Embryol (Berl) 156:1–28

Edwards JA (1968) The external development of the rabbit and rat embryo. Adv Teratol 3:239–263

Ferron RR, Miller RS, McNutty WP (1976) Estimation of fetal age and weight from radiographic skull diameters in the rhesus monkey (*Macaca mulatta*). J Med Primatol 5:41–48

Festing MFW (1976) Hamsters. In: The UFAW handbook on the care and management of laboratory animals. Churchill Livingstone, Edingburgh, pp 248–262

Fulton JT (1957) Closure of the human palate in embryo. Am J Obstet Gynecol 74:179–182

Graves AP (1945) Development of the golden hamster, *Cricetulus auratus*, during the first nine days. Am J Anat 77:219–251

Greene RM, Pratt RM (1976) Developmental aspects of secondary palate formation. J Embryol Exp Morphol 36:225–245

Gribnau AAM (1975) Immunologic pregnancy test in the rhesus monkey (*Macaca mulatta*). J Med Primatol 4:65–69

Gribnau AAM, Lammers GJ (1976) The preparation of graphical and three-dimensional reconstructions of the developing central nervous system. Acta Morphol Neerl Scand 14:1–18

Hendrickx AG (1971) Embryology of the baboon. University of Chicago Press, Chicago

Hendrickx AG (1972a) Early development of the embryo in non-human primates and man. Acta Endocrinol [Suppl] (Copenh) 166:103–130

Hendrickx AG (1972b) A comparison of temporal factors in the embryological development of man, old world monkeys and galagos, and craniofacial malformations induced by thalidomide and triamcinolone. In: Goldsmith EJ, Moor-Jankowski J (eds) Medical primatology 1972. Proc. 3rd Conf. exp. Med. Surg. Primates, Lyon 1972, part III. Karger, Basel, pp 259–269

Hendrickx AG, Houston ML (1971) Prenatal and postnatal development. In: Hafez ESE (ed) Comparative reproduction of non-human primates. Thomas, Springfield, pp 334–381

Hendrickx AG, Sawyer RH (1975) Embryology of the rhesus monkey. In: Bourne GH (ed) The rhesus monkey: vol II, Management, reproduction and pathology. Academic Press, New York, pp 141–169

Hendrickx AG, Sawyer RH, Lasley BL, Barnes RD (1975) Comparison of developmental stages in primates with a note on the detection of ovulation. In: Perkins FT, O'Donoghue PN (eds) Breeding simians for developmental biology. Laboratory Animals Ltd, London (Laboratory Animal Handbooks, vol 6, pp 305–315)

Hertig AT, Rock J, Adams EC (1956) A description of 34 human ova within the first 17 days of development. Am J Anat 98:435–493

Heuser CH, Corner GW (1957) Developmental horizons in human embryos. Description of age group X, 4 to 12 somites. Contrib Embryol Carneg Instn 36:29–39

Heuser CH, Streeter GL (1941) Development of the macaque embryo. Contrib Embryol Carneg Instn 29:15–55

Hughes V, Furstman L, Bernick S (1967) Prenatal development of the rat palate. J Dent Res 46:373–379

Jackson CG (1976) Prenatal development of the eye in the golden hamster. Am J Anat 146: 303–322

Jacobson AG, Miyamoto DM, Mai SH (1979) Rathke's pouch morphogenesis in the chick embryo. J Exp Zool 207:351–390

Kerr GR, Wallace JH, Chesney CF, Waisman HA (1972) Growth and development of the fetal rhesus monkey. III. Maturation and linear growth of the skull and appendicular skeleton. Growth 36:59–76

Koch WE, Smiley GR (1973) An analysis of "fusion" during secondary palate formation. Anat Rec 175:361

Kuhn HJ, Schwaier A (1973) Implantation, early placentation, and the chronology of embryogenesis in *Tupaia belangeri.* Z Anat Entwickl-Gesch 142:315–340

Lammers GJ, Gribnau AAM, ten Donkelaar HJ (1980) Neurogenesis in the basal forebrain in the Chinese hamster (*Cricetulus griseus*). II. Site of neuron origin: morphogenesis of the ventricular ridges. Anat Embryol (Berl) 158:193–211

Lemire RJ, Loeser JD, Leech RW, Alvord EC (1975) Normal and abnormal development of the human nervous system. Harper & Row, Hagerstown Maryland New York Evanston San Francisco London

Moore KL (1973) The developing human. Clinically oriented embryology. Saunders, Philadelphia London Toronto

Nishimura H, Takano K, Tanimura T, Yasuda M (1968) Normal and abnormal development of human embryos: first report of the analysis of 1,123 intact embryos. Teratology 1:281–290

Olivier G, Pineau H (1962) Horizons de Streeter et âge embryonnaire. CR Ass Anat 47:573–576

O'Rahilly R (1963) The early development of the otic vesicle in staged human embryos. J Embryol Exp Morphol 11:741–755

O'Rahilly R (1966) The early development of the eye in staged human embryos. Contrib Embryol Carneg Instn 38:1–42

O'Rahilly R (1967) The early development of the nasal pit in staged human embryos. Anat Rec 157:380

O'Rahilly R (1968) The development of the epiphysis cerebri and the subcommissural complex in staged human embryos. Anat Rec 160:448–489

O'Rahilly R (1972) Guide to the staging of human embryos. Anat Anz 130:556–559

O'Rahilly R (1973a) Developmental stages in human embryos, including a survey of the Carnegie collection. Part A: Embryos of the first three weeks (stages 1 to 9). Carnegie Institution of Washington Public, Baltimore, p 631

O'Rahilly R (1973b) The early development of the hypophysis cerebri in staged human embryos. Anat Rec 175:511

O'Rahilly R (1975) The prenatal development of the human eye. Exp Eye Res 21:93–112

O'Rahilly R (1978) The timing and sequence of events in the development of the human digestive system and associated structures during the embryonic period proper. Anat Embryol (Berl) 153:123–136

O'Rahilly R (1979) Early human development and the chief sources of information on staged human embryos. Eur J Obstet Gynecol Reprod Biol 9/4:273–280

O'Rahilly R, Gardner E (1971) The timing and sequence of events in the development of the human nervous system during the embryonic period proper. Z Anat Entwickl-Gesch 134: 1–12

O'Rahilly R, Gardner E (1974) Les stades de development de l'embryo humain. Bull Assoc Anat 58/160:177–182

O'Rahilly R, Gardner E (1975) The timing and sequence of events in the development of the limbs in the human embryo. Anat Embryol (Berl) 148:1–23

O'Rahilly R, Gardner E (1977) The developmental anatomy and histology of the human central nervous system. In: Vinken PJ, Bruyn GW, Myrianthopoulos NC (eds) Handbook of clinical neurology, vol 30. North-Holland, Amsterdam, pp 15–40

O'Rahilly R, Gardner E (1979) The initial development of the human brain. Acta Anat 104:123–133

O'Rahilly R, Meyer DB (1979) The timing and sequence of events in the development of the human vertebral column during the embryonic period proper. Anat Embryol (Berl) 157: 167–176

O'Rahilly R, Müller F (1978) A model of the pancreas to illustrate its development. Acta Anat 100:380–385

O'Rahilly R, Gardner E, Gray DJ (1956) The ectodermal thickening and ridge in the limbs of staged human embryos. J Embryol Exp Morphol 4:254–264

O'Rahilly R, Gray DJ, Gardner E (1957) Chondrification in the hands and feet of staged human embryos. Contrib Embryol Carneg Instn 36:183–192

Otis EM, Brent R (1954) Equivalent ages in mouse and human embryos. Anat Rec 120:33–64

Pei YF, Rhodin JAG (1970) The prenatal development of the mouse eye. Anat Rec 168:105–126

Phillips IR (1976) The embryology of the common marmoset (*Callithrix jacchus*). Adv Anat Embryol Cell Biol 52: Fasc 5

Romeis B (1968) Mikroskopische Technik. Oldenbourg, München

Rugh R (1968) The mouse. Its reproduction and development. Burgess, Minneapolis

Schultz AH (1937) Fetal growth and development of the rhesus monkey. Contrib Embryol Carneg Instn 26:71–97

Scott JP (1937) The embryology of the Guinea pig. Am J Anat 60:397–432

Smiley GR, Koch WE (1971) Fine structure of mouse secondary palate development in vitro. J Dent Res 50:1671–1677

Steffek AJ, Verrusio AC, King CTG (1968) The histology of palatal closure in the rhesus monkey (*Macaca mulatta*). Teratology 1:425–430

Streeter GL (1922) Development of the auricle in the human embryo. Contrib Embryol Carneg Instn 14:111–138

Streeter GL (1942) Developmental horizons in human embryos. Description of age group XI, 13–20 somites, and age group XII, 21 to 29 somites. Contrib Embryol Carneg Instn 30: 211–245

Streeter GL (1945) Developmental horizons in human embryos. Description of age group XIII, embryos about 4 or 5 millimetres long and age group XIV, period of indentation of the lens vesicle. Contrib Embryol Carneg Instn 31:27–63

Streeter GL (1948) Developmental horizons in human embryos. Description of age groups XV, XVI, XVII and XVIII, being the third issue of a survey of the Carnegie collection. Contrib Embryol Carneg Instn 32:133–203

Streeter GL (1951) Developmental horizons in human embryos. Description of age groups XIX, XX, XXI and XXIII, being the fifth issue of a survey of the Carnegie collection. Contrib Embryol Carneg Instn 34:165–196

Tanimura T, Tanioka Y (1975) Comparison of embryonic and fetal development in man and rhesus monkey. In: Perkins FT, O'Donoghue PN (eds) Breeding simians for developmental biology. Laboratory Animals Ltd, London (Laboratory Animal Handbooks, vol 6, pp 205–233)

Theiler K (1972) The house mouse. Development and normal stages from fertilization to 4 weeks of age. Springer, Berlin Heidelberg New York

Tullner WW (1968) Urinary chorionic gonadotropin excretion in the monkey (*Macaca mulatta*) early phase. Endocrinology 82:875–875

Wagenen G van, Asling CW (1964) Ossification in the fetal monkey (*Macaca mulatta*). Estimation of age and progress of gestation by roentgenography. Am J Anat 114:107–132

Wagenen G van, Simpson M (1965) Embryology of the ovary and testis *Homo sapiens* and *Macaca mulatta*. Yale University Press, New Haven

Wagenen G van, Catchpole HR, Negri J, Butzko D (1965) Growth of the fetus and placenta of the monkey (*Macaca mulatta*). Am J Physiol Anthropol 23:23–33

Walker BE, Fraser FC (1956) Closure of the secondary palate in three strains of mice. J Embryol Exp Morphol 4:176–189

Wilson JG (1973) Environment and birth defects. Academic Press, New York

Wislocki GB, Streeter GL (1938) On the placentation of the macaque (*Macaca mulatta*) from the time of implantation until the formation of the definitive placenta. Contrib Embryol Carneg Instn 27:1–66

Zeiler KB, Weinstein S, Gibson RD (1964) A study of the morphology and the time of closure of the palate in the albino rat. Arch Oral Biol 9:545–554

Ziehen T (1906) Morphogenie des Zentralnervensystems der Säugetiere. In: Hertwig O (ed) Handbuch der vergleichenden und experimentellen Entwicklungslehre der Wirbeltiere. Fischer, Jena

Subject Index

Springer-Verlag Berlin Heidelberg New York

H. Stephan, G. Baron, W. Schwerdtfeger

The Brain of the Common Marmoset (Callithrix jacchus)

A Stereotaxic Atlas

1980. 5 figures, 3 tables, 73 plates. V, 91 pages
ISBN 3-540-09782-1

Contents: Introduction. – Material. – Zero Coordinates. – Reference Coordinates. – Histological Procedures. – Shrinkage Factors. – Variability. – Standardization Proposals. – Photomicrographs for Atlas. – Presentation and Nomenclature. – References. – Index.

As anthropoid or simian primates, marmosets are close to man on the evolutionary scale. In contrast to other simians, the care and breeding of marmosets is relatively easy and their reproduction rate demonstrably higher. These characteristics make marmosets of great interest for primate research, especially as the supply of simians from the wild becomes increasingly limited.

This atlas presents 73 photomicrographs of successive serial sections from the brains of common marmosets. Taken at 0.5 mm intervals, the sections are stained for both cells and nerve fibers and shown on facing pages for comparison. Stereotactic coordinates are labelled in detail and methods for their standardization discussed.

The accuracy of the stereotactic descriptions provided here will greatly aid researchers using marmoset brains for neurophysiologic, neuroanatomic, neurochemical and behavioral investigations.

Springer-Verlag
Berlin
Heidelberg
New York